Integral Matrices

This is Volume 45 in
PURE AND APPLIED MATHEMATICS
A Series of Monographs and Textbooks
Editors: PAUL A. SMITH AND SAMUEL EILENBERG
A complete list of titles in this series appears at the end of this volume.

INTEGRAL MATRICES

MORRIS NEWMAN

U. S. Department of Commerce
National Bureau of Standards
Washington, D. C.

ACADEMIC PRESS New York and London 1972

ACADEMIC PRESS, INC.
111 Fifth Avenue, New York, New York 10003

United Kingdom Edition published by
ACADEMIC PRESS, INC. (LONDON) LTD.
24/28 Oval Road, London NW1

LIBRARY OF CONGRESS CATALOG CARD NUMBER: 72-182628

AMS (MOS) 1970 Subject Classifications: 15A21, 15A36, 20H05, 10C05

PRINTED IN THE UNITED STATES OF AMERICA

Contents

Chapter VII **Groups of Matrices**

Chapter XI **Quadratic Forms**

Preface

Integral matrices, or matrices over rings, is a huge subject which extends into many different areas of mathematics. For example, the entire theory of finite groups could be placed in this category. It goes without saying then that the selection of material for a book with such a title must necessarily be limited, and a matter of personal preference.

Since it is the matrix theory with its many applications and not abstract ring theory which is the primary concern of this book, the discussion for the most part has been limited to matrices over principal ideal rings. Chapter I provides the necessary background material for this subject. Chapters II, III, IV, and V are concerned with the various kinds of equivalence relationships that may be defined for matrices. Among other things, the classical theory of matrices over a field is developed in these chapters. Chapter VI is a self-contained presentation of the Minkowski geometry of numbers. Its principal purpose is to make available the powerful methods of the subject to the theory of quadratic forms. Chapter VII presents the theory of matrix groups over principal ideal rings, with special attention to the congruence groups, which have assumed new importance because of the recent work of J. L. Mennicke, H. Bass, M. Lazard, and J.-P. Serre. Chapter

VIII discusses the classical modular group and certain generalizations, such as the free product of two cyclic groups. Chapter IX presents the elements of the theory of group representations, with applications to finite integral matrix groups. Chapter X is devoted to the study of the group matrix of a cyclic group, and Chapter XI to the theory of quadratic forms in n dimensions.

Since matrix theory is the basic discipline of this book, a knowledge of the elementary properties of vector spaces over fields and matrices whose elements lie in a field is assumed. The reader must have no difficulty in handling partitioned matrices, and must be familiar with the body of formal theorems relating to products of partitioned matrices, determinants of partitioned matrices, etc. Furthermore, elementary facts about rank, systems of linear equations, eigenvalues and eigenvectors, adjoints, characteristic and minimal polynomials, Kronecker products, etc. are used freely. These are not proved here, but may be found in any of the classical texts on matrix theory, for example, MacDuffee [28] or Marcus and Minc [31].

An understanding of elementary group theory is also required, and the reader should be familiar with this subject. Good general references are Hall [13] or Zassenhaus [72]. At one point, however, a difficult theorem from group theory must be used. This occurs in Chapter VIII, where the Kurosh subgroup theorem forms the basis for the discussion of the classical modular group. Good expositions of this theorem may be found in Hall [13] or Kurosh [23]. The alternative was to develop the subject from the geometric viewpoint; but this leads to difficulties of another type. There is no easy alternative.

Free use of the concepts and results of elementary number theory is made everywhere, and the reader must be thoroughly familiar with this subject. Good elementary texts are Hardy and Wright [15] or Landau [24]. In addition the classical theory of algebraic numbers is used from time to time, and the reader should also be familiar with the rudiments of this subject. An elegant presentation may be found in Hecke's book [16]. For simple expositions see Newman [52] or Pollard [59].

A one- or two-semester graduate course could be fashioned from this book very easily. The first five chapters would form the core of the course and could be covered in a semester. The next semester could be devoted to special topics selected from the remaining chapters at the discretion of the instructor. Enough open problems and suggestions for research are mentioned to provide direction for further reading or for individual research.

Acknowledgments

The orgin of this book was a series of lectures delivered a number of years ago at the University of British Columbia, at the invitation of Marvin Marcus. Notes were taken then which served as a skeletal outline for the final work. Much of the author's personal research was done jointly with Olga Taussky-Todd, and a good part of the book is devoted to an exposition of this joint work and to work done solely by her. Others to whom I am grateful for work done jointly, or for their own stimulating research, are M. Hall, M. I. Knopp, J. Lehner, I. Reiner, H. Ryser, and J. R. Smart. Finally, my deepest thanks go to Mrs. Doris Burrell for her painstaking efforts in typing the manuscript.

Chapter I

Background Material on Rings

Although classical matrix theory is concerned with matrices over the ring of integers Z or over a field, there are definite advantages to be gained by adopting a more general approach and studying matrices over any principal ideal ring. The properties of principal ideal rings do not differ markedly from the properties of Z, and we will develop these in what follows.

1. Principal ideal rings

A *principal ideal ring* R is a commutative ring with no zero divisors and a unit element 1 such that every ideal is principal; that is, such that every ideal consists of the totality of multiples of a fixed element. It is not actually necessary to assume that R contains a unit element, since this is derivable from the other hypotheses. Examples of principal ideal rings are:

(1) The ring of integers Z.
(2) Any field F.

(3) The ring of polynomials $F[x]$ in a single variable x, with coefficients from a field F.

In what follows R will always denote a principal ideal ring, and F a field.

2. Units

Let S be any commutative ring with a unit element 1. If α is any element of S such that $\alpha\beta = 1$ for some other element β of S, then α is said to be a *unit* of S. The totality of units of S forms a multiplicative group, which we denote by S'. In Z, for example, the units are $1, -1$ and, in any field F, the units are the nonzero elements. In $F[x]$, the units are the nonzero elements of F.

An element α of S is said to be an *associate* of an element β of S, if $\alpha = \beta\varepsilon$, where ε belongs to S'. It is readily verified that this is an equivalence relationship over S, so that S decomposes into equivalence classes with respect to this relationship. The class containing 0 (the zero of S) consists of 0 alone, and the class containing 1 consists of S'. A set of elements of S, one from each equivalence class, is said to be a *complete set of non associates*. This will always be chosen so that 1 is the representative of its class.

We now return to R.

3. Divisibility

If α, β are any elements of R and $\beta \neq 0$, we say that β *divides* α (written $\beta \mid \alpha$) if $\alpha = \beta\gamma$, for some element γ of R. We readily verify that $\beta \mid \alpha$ if and only if $(\beta) \supset (\alpha)$, where (γ) is the principal ideal consisting of the totality of multiples of γ, for any element γ of R. The units are trivial divisors of any element of R. Also we note that if α, β are nonzero elements of R, then $(\alpha) = (\beta)$ if and only if α and β are associates.

Let α, β be any elements of R which are not both 0. A *greatest common divisor* of α and β is an element δ of R such that $\delta \mid \alpha, \delta \mid \beta$; and if γ is any element of R which divides both α and β, then $\gamma \mid \delta$. It is immediate that any pair of greatest common divisors of α and β

are associates. The fact that R is a principal ideal ring guarantees the existence of at least one greatest common divisor of α and β. We let (α, β) denote the ideal consisting of the totality of elements $\mu\alpha + \nu\beta$, where μ, ν run over R independently. We have

Theorem I.1. Let α, β be any elements of R which are not both 0. Let $(\alpha, \beta) = (\delta)$, say. Then δ is a greatest common divisor of α and β.

Proof. Certainly, $\delta \neq 0$. Since α and β each belong to (δ), $\delta \mid \alpha$, and $\delta \mid \beta$. Furthermore, elements μ, ν of R exist such that $\delta = \mu\alpha + \nu\beta$. Hence if γ is any divisor of α and β, then $\gamma \mid \delta$. Thus δ is a greatest common divisor of α and β and the proof is complete.

The preceding discussion may be generalized in obvious fashion. If $\alpha_1, \alpha_2, \ldots, \alpha_n$ are elements of R which are not all 0, the ideal $(\alpha_1, \alpha_2, \ldots, \alpha_n)$ consisting of the totality of elements $\mu_1\alpha_1 + \mu_2\alpha_2 + \cdots + \mu_n\alpha_n$, where $\mu_1, \mu_2, \ldots, \mu_n$ run over R independently, is principal and equal to (δ), say; and δ is a greatest common divisor of $\alpha_1, \alpha_2, \ldots, \alpha_n$. We write

$$(\alpha_1, \alpha_2, \ldots, \alpha_n) = \delta$$

The conflict of notation will cause no confusion. It will be clear from the context whether we are referring to an ideal or to a ring element. Notice also that δ may be replaced by any of its associates. If δ is a unit we say that $\alpha_1, \alpha_2, \ldots, \alpha_n$ are *relatively prime* and write

$$(\alpha_1, \alpha_2, \ldots, \alpha_n) = 1$$

If we restrict δ to lie in some complete set of nonassociates, then δ is uniquely defined. We shall do this in what follows.

4. Congruence and norms

Let μ be any nonzero element of R. We say that α *is congruent to* β *modulo* μ (written $\alpha \equiv \beta \bmod \mu$) if μ divides $\alpha - \beta$; or what is the same thing, if $\alpha - \beta$ belongs to (μ). Once again, this is an equivalence relationship over R. A set of elements of R, one from each equivalence class, is said to be a *complete set of residues modulo* μ. In order to complete the definition, a complete set of residues modulo 0 will be taken to be the whole ring R, and $\alpha \equiv \beta \bmod 0$ to mean that $\alpha = \beta$.

The number of equivalence classes (which may be infinite) will be denoted by $N(\mu)$ (the *norm* of μ), and is just the order of the quotient ring $R/(\mu)$. For example, if $R = Z$, then $N(\mu) = |\mu|$, and if R is a field, then $N(\mu) = 1$.

The following result will be of later use:

Theorem I.2. Suppose that μ, ν are nonzero elements of R such that $N(\mu)$, $N(\nu)$ are finite. Then $N(\mu\nu)$ is also finite, and $N(\mu\nu) = N(\mu)N(\nu)$.

Proof. Let μ_i, $1 \leq i \leq m = N(\mu)$, be a complete set of residues modulo μ, and ν_j, $1 \leq j \leq n = N(\nu)$, a complete set of residues modulo ν. We will show that the mn elements of R

$$\rho_{ij} = \mu_i + \mu\nu_j, \qquad 1 \leq i \leq m, \quad 1 \leq j \leq n$$

is a complete set of residues modulo $\mu\nu$. Suppose that $\rho_{ij} \equiv \rho_{kl} \bmod \mu\nu$, where $1 \leq i, k \leq m$, $1 \leq j, l \leq n$, Then certainly $\rho_{ij} \equiv \rho_{kl} \bmod \mu$, so that $\mu_i \equiv \mu_k \bmod \mu$. Hence $i = k$, and so $\mu\nu_j \equiv \mu\nu_l \bmod \mu\nu$, so that $\nu_j \equiv \nu_l \bmod \nu$. Hence $j = l$, and no two of the elements ρ_{ij} are congruent modulo $\mu\nu$. Now let α be any element of R. Then for some i such that $1 \leq i \leq m$, $\alpha \equiv \mu_i \bmod \mu$, so that $\alpha = \mu_i + \mu\beta$ for some element β of R. Also for some j such that $1 \leq j \leq n$, $\beta \equiv \nu_j \bmod \nu$. It follows that $\alpha \equiv \mu_i + \mu\nu_j \bmod \mu\nu$. This completes the proof.

5. The ascending chain condition

Perhaps the most important single fact about any principal ideal ring is that it must satisfy the ascending chain condition: that is, if

$$A_1 \subset A_2 \subset \cdots$$

is any infinite sequence of ideals, then for some positive integer k,

$$A_k = A_{k+1} = \cdots$$

We state this as a theorem.

Theorem I.3. Any principal ideal ring R satisfies the ascending chain condition.

Proof. Let

$$A_1 \subset A_2 \subset \cdots$$

be an infinite sequence of ideals of R, and let $A = \bigcup_{i=1}^{\infty} A_i$ be their union. Then A is certainly an ideal. Since R is a principal ideal ring, we have that $A = (\alpha)$ for some element α of R. Now α must belong to A_k for some positive integer k, and so to A_l for all $l \geq k$. Hence $A \subset A_l$, $l \geq k$. But trivially $A \supset A_l$, $l \geq k$. It follows that

$$A = A_k = A_{k+1} = \cdots$$

This completes the proof.

6. Unique factorization

One of the most interesting conclusions to be drawn about principal ideal rings is that such a ring admits a theory of unique factorization (or is *Gaussian*). With the demonstration of this fact in mind, we say that a nonzero element π of a principal ideal ring R is a *prime* if it is not a unit and if its divisors are either units or associates of π. We now use the ascending chain condition to prove

Lemma I.1. Every nonzero element of R which is not a unit is the product of finitely many primes.

Proof. Let α be any nonzero element of R which is not a unit. We first show that α is divisible by at least one prime. If α is a prime, we are done. If not, α has a factorization $\alpha = \alpha_1\beta_1$, where neither α_1 nor β_1 is a unit. If α_1 is a prime, we are done. If not, α_1 has a factorization $\alpha_1 = \alpha_2\beta_2$, where neither α_2 nor β_2 is a unit. This process must terminate, since otherwise we obtain an ascending sequence of ideals $(\alpha_1) \subset (\alpha_2) \subset \cdots$ such that $\alpha_k = \alpha_{k+1}\beta_{k+1}$ and β_{k+1} is not a unit, $k \geq 1$. But then the ascending chain condition implies that for some positive integer l, $(\alpha_l) = (\alpha_{l+1}) = \cdots$. Thus α_l and α_{l+1} are associates, which implies that β_{l+1} is a unit, a contradiction. It follows that α is divisible by at least one prime.

Now write $\alpha = \pi_1\beta_1$, where π_1 is a prime. If β_1 is not a unit, write $\beta_1 = \pi_2\beta_2$, where π_2 is a prime. Once again, this process must terminate, since otherwise we have that for every $k \geq 1$, $\beta_k = \pi_{k+1}\beta_{k+1}$, where π_{k+1} is a prime. The ascending chain condition applied to the chain $(\beta_1) \subset (\beta_2) \subset \cdots$ now implies that for some positive integer l, $(\beta_l) = (\beta_{l+1}) = \cdots$. But then β_l and β_{l+1} are associates, implying that π_{l+1} is a unit, which is a contradiction. This completes the proof.

We next prove

Lemma I.2. Let α, β be elements of R, and let π be a prime. Then if $\pi \mid \alpha\beta$, $\pi \mid \alpha$ or $\pi \mid \beta$.

Proof. Suppose that π does not divide α. Consider the ideal $(\pi, \alpha) = (\delta)$, say. Since $\delta \mid \pi$ and π is a prime, δ is a unit or an associate of π. But δ cannot be an associate of π, since δ also divides α. It follows that δ must be a unit. Thus π and α are relatively prime, and elements μ, ν of R exist such that $\mu\pi + \nu\alpha = 1$. Thus $\mu\pi\beta + \nu\alpha\beta = \beta$, and so $\pi \mid \beta$, since $\pi \mid \alpha\beta$. This completes the proof.

Notice that this lemma implies that if π divides the product of any finite number of elements of R, then π divides at least one of them.

We are now in a position to prove

Theorem I.4. Any nonzero element of R which is not a unit has a unique expression as the product of finitely many primes, apart from unit factors and order.

Proof. Let α be a nonzero element of R which is not a unit. By Lemma I.1, α may be written as $\alpha = \pi_1\pi_2 \cdots \pi_r$, where $\pi_1, \pi_2, \ldots, \pi_r$ are primes. Let $\alpha = \pi_1'\pi_2' \cdots \pi_s'$ be any other factorization of this type, and assume (as we may) that $r \leq s$. Since $\pi_1 \mid \pi_1'\pi_2' \cdots \pi_s'$, Lemma I.2 implies that $\pi_1 \mid \pi_k'$ for some k such that $1 \leq k \leq s$. After a suitable rearrangement, we may assume that $k = 1$, so that $\pi_1 \mid \pi_1'$. Since π_1' is a prime, π_1 and π_1' must be associates. Canceling π_1, we get

$$\pi_2 \cdots \pi_r = \varepsilon_1\pi_2' \cdots \pi_s'$$

where ε_1 is a unit. Repeating the argument we find that after suitable renumbering π_2 and π_2' are associates, π_3 and π_3' are associates, $\ldots$, π_r and π_r' are associates. Hence r must equal s, for otherwise the product $\pi_{r+1}' \cdots \pi_s'$ would contain at least one prime factor and would also be a unit. This completes the proof.

A useful corollary of this theorem is the following:

Corollary I.1. Let R be a principal ideal ring. Then it is always possible to choose a complete set of nonassociates for R which is multiplicatively closed.

Proof. Let P be any complete set of nonassociated primes of R; that is, no two elements of P are associated, and any prime of R is the associate of some element of P. Define M as the set consisting of 0, 1, and all products of finitely many elements of P, allowing repetition. Then certainly M is multiplicatively closed; and in view of Theorem I.4, M is a complete set of nonassociates.

7. Euclidean rings

We turn now to a discussion of euclidean rings. It will transpire that every euclidean ring is itself a principal ideal ring, but not conversely. That is, there do exist principal ideal rings which are not euclidean. An interesting discussion of this point is to be found in [36].

A commutative ring E with no divisors of zero and a unit element 1 is a *euclidean ring*, if there is a nonnegative integer-valued function v defined over the nonzero elements of E with the following properties:

(4) $v(\alpha\beta) \geq v(\alpha)$ for all nonzero elements α, β of E.
(5) If α, β are arbitrary elements of E and $\beta \neq 0$, then elements τ, ρ of E exist such that $\alpha = \tau\beta + \rho$, and either $\rho = 0$, or if $\rho \neq 0$, then $v(\rho) < v(\beta)$.

Once again we remark parenthetically that it is not necessary to assume that E possesses a unit element, since this is derivable from the other hypotheses.

We now prove

Theorem I.5. A euclidean ring is a principal ideal ring.

Proof. Let A be any ideal of the euclidean ring E. We may assume that A is not the zero ideal. From the nonzero elements of A choose one, say β, with minimal v. Then $A = (\beta)$. For suppose that α is any element of A. By (5) above, elements τ, ρ of E exist such that $\alpha = \tau\beta + \rho$, and $v(\rho) < v(\beta)$ if $\rho \neq 0$. Since $\rho = \alpha - \tau\beta$ belongs to A, this is not possible. Hence $\rho = 0$ and $\alpha = \tau\beta$, which implies that $A \subset (\beta)$. But trivially, $A \supset (\beta)$, and so $A = (\beta)$. This completes the proof.

Some examples of euclidean rings are the ring of integers Z, with $v(\alpha) = N(\alpha) = |\alpha|$; any field F, with $v(\alpha) = 1$; and the ring $F[x]$ of polynomials in a single variable x over F, with $v(\alpha) = \deg(\alpha)$.

It is useful to investigate the function v a little further. Let α, β be nonzero elements of E and suppose that $v(\alpha\beta) = v(\alpha)$. Consider the ideal $A = (\alpha)$. By the method of proof employed in Theorem I.5, we see that A must also equal $(\alpha\beta)$, since $v(\alpha)$ is the least value that v assumes for any nonzero element of A, and any nonzero element of A for which v is minimal must generate A. It follows that α and $\alpha\beta$ are associates, so that β is a unit.

Conversely, suppose that β is a unit, and that α is any nonzero element of E. Then certainly $v(\alpha\beta) \geq v(\alpha)$. But also (since β is a unit)

$$v(\alpha) = v(\alpha\beta \cdot \beta^{-1}) \geq v(\alpha\beta)$$

Thus $v(\alpha\beta) = v(\alpha)$. We have shown therefore that equality in (4) holds if and only if β is a unit.

A noteworthy special case is obtained by choosing $\alpha = 1$. The conclusion is that β is a unit if and only if $v(\beta) = v(1)$.

8. The Chinese remainder theorem

Practically the whole of elementary number theory, including the theory of congruences, goes over unchanged for any principal ideal ring. For example, the Chinese remainder theorem holds in any principal ideal ring R. This states that if $\mu_1, \mu_2, \ldots, \mu_n$ are pairwise relatively prime elements of R, and if $\alpha_1, \alpha_2, \ldots, \alpha_n$ are arbitrary elements of R, then the system of congruences

$$x \equiv \alpha_i \bmod \mu_i, \qquad 1 \leq i \leq n$$

always has a solution x in R and this solution is unique modulo $\mu = \mu_1\mu_2 \cdots \mu_n$. The proof is precisely the same as in the classical case and is as follows: Set $\tau_i = \mu/\mu_i$, $1 \leq i \leq n$. Then $(\tau_i, \mu_i) = 1$, since the elements μ_i are pairwise relatively prime. Hence an element μ_i' of R can be determined so that $\tau_i\mu_i' \equiv 1 \bmod \mu_i$, $1 \leq i \leq n$. But then

$$x = \sum_{i=1}^{n} \tau_i\mu_i'\alpha_i$$

is certainly a solution of the system. Furthermore any two solutions x, y must satisfy

$$x - y \equiv 0 \bmod \mu_i, \qquad 1 \leq 1 \leq n$$

so that

$$x - y \equiv 0 \bmod \mu, \qquad x \equiv y \bmod \mu,$$

since the μ_i are pairwise relatively prime.

Such results will be used freely and with no further comment in what follows.

EXERCISES AND PROBLEMS

In the following, R is a principal ideal ring.

1. Prove that a commutative ring such that every ideal is principal automatically contains a unit element.

2. Give an example of a ring with elements α, β, γ such that $\alpha\beta = \alpha\gamma = 1$, but $\beta \neq \gamma$.

3. Prove that R is a field if and only if R contains no primes.

4. Prove that R is a field if and only if R satisfies the descending chain condition.

5. Prove that R is a field if and only if a complete set of nonassociates contains only finitely many elements.

6. Prove that if α, β, γ are elements of R, then $(\alpha, \beta, \gamma) = ((\alpha, \beta), \gamma)$, where ( ) denotes greatest common divisor.

7. Suppose that $N(\mu)$ is finite for all nonzero μ of R. Define $\varphi(\mu)$ as the number of elements α in a complete set of residues modulo μ such that $(\alpha, \mu) = 1$. Show that

(i) $$\varphi(\mu\nu) = \varphi(\mu)\varphi(\nu), \qquad (\mu, \nu) = 1$$

(ii) $$\varphi(\pi^n) = N(\pi)^n(1 - 1/N(\pi)), \qquad \pi \text{ prime}$$

8. Prove that the ring of Gaussian integers $Z[i]$ is Euclidean, if v is defined as $v(\alpha) = \alpha\bar{\alpha} = |\alpha|^2$ for any element α of $Z[i]$. What are the units of this ring?

9. Find all the units of $Z[\sqrt{2}]$.

10. Let Q be the rational field, and $Q\{x\}$ the ring of formal power series in x with coefficients from Q. Prove the following:

(i) The element $a_0 + a_1x + \cdots$ of $Q\{x\}$ is a unit if and only if $a_0 \neq 0$, and hence every element of $Q\{x\}$ is an associate of x^k for some $k \geq 0$.
(ii) $Q\{x\}$ is a principal ideal ring.
(iii) Apart from unit multipliers, the only prime of $Q\{x\}$ is x.

This furnishes an example of a principal ideal ring with only finitely many nonassociated primes. Is $Q\{x\}$ euclidean?

11. Prove that if μ is a nonzero element of R which is not a unit, then $R/(\mu)$ is a field if and only if μ is a prime.

Chapter II

Equivalence

1. Definition of matrix ring S_n over S

Let S be a commutative ring with unit element 1. Let S_n denote the ring of $n \times n$ matrices over S. Then if $n > 1$, S_n is no longer commutative, possesses zero divisors, but does have a unit element

$$I = I_n = (\delta_{ij})$$

where δ_{ij} is the Kronecker symbol. It is known that every ideal of S_n is necessarily of the form A_n, where A is an ideal of S, and A_n is the ring of $n \times n$ matrices over A. A similar result will be proved later for left ideals, in the case when S is a principal ideal ring.

2. Units of S_n

The units of S_n are obtained as follows: The matrix A of S_n is a unit if and only if $AB = I$ for some matrix B of S_n. If $AB = I$, then $\det(A)\det(B) = 1$, so that $\det(A)$ is a unit of S. If $\det(A)$ is a unit of S then the matrix

$$B = (\det(A))^{-1}A^{\text{adj}}$$

belongs to S_n and satisfies $AB = BA = I$, so that A is a unit of S_n. Hence A is a unit of S_n if and only if $\det(A)$ is a unit of S. The matrices of S_n which are units will be referred to as *unimodular* matrices, and of course form a multiplicative group, which we denote by S_n' in conformity with previous notation.

We now choose $S = R$, a principal ideal ring. Some examples illustrating the remarks above follow.

(1) $R = Z$, the ring of integers. The unimodular matrices of R_n are those of determinant ± 1.
(2) $R = F$, a field. The unimodular matrices of R_n are the nonsingular matrices.
(3) $R = F[x]$. The unimodular matrices of R_n are those whose determinant is a nonzero element of F.

In the previous chapter we developed a complete theory of factorization for the elements of R. In this chapter we set ourselves a similar object, that of finding canonical forms for the elements A of R_n under transformations of the form $A \to UA$, $A \to AV$, $A \to UAV$, where U, V are unimodular matrices of R_n.

3. Definition of equivalence

Let A, B be matrices of R_n. We say that B is *left equivalent* to A (written $B \tilde{L} A$) if $B = UA$ for some unimodular matrix of R_n; B is *right equivalent* to A (written $B \tilde{R} A$) if $B = AV$ for some unimodular matrix V of R_n; and B is *equivalent* to A (written $B \tilde{E} A$) if $B = UAV$ for some unimodular matrices U, V of R_n. These are all equivalence relationships.

4. Elementary row operations

We consider the following *elementary row operations* performed on a matrix A of R_n:

(4) The interchange of two rows.
(5) The multiplication of a row by a unit of R.

(6) The addition of μ times one row to another row, where μ belongs to R.

Each of these operations corresponds to multiplication of A on the left by a suitable element of R_n'. Thus,

$$\begin{bmatrix} 0 & 1 \\ 1 & 0 \end{bmatrix} \dotplus I_{n-2}$$

interchanges rows 1 and 2;

$$\mathrm{diag}(\varepsilon, 1, \ldots, 1)$$

multiplies row 1 by the unit ε of R; and

$$\begin{bmatrix} 1 & \mu \\ 0 & 1 \end{bmatrix} \dotplus I_{n-2}$$

adds μ times row 2 to row 1, where μ belongs to R.

Matrices such as the above, which effect elementary operations, will be called *elementary* matrices. They play an important role in the classification of matrices with respect to the equivalence relationships defined above. Elementary column operations are defined in entirely analogous fashion.

5. Completion to a unimodular matrix

The theorem that follows is basic to the subject.

Theorem II.1. Let $\alpha_1, \alpha_2, \ldots, \alpha_n$ be elements of R, and let $\delta_n = (\alpha_1, \alpha_2, \ldots, \alpha_n)$ be their greatest common divisor. Then there is a matrix D_n of R_n with first row $[\alpha_1, \alpha_2, \ldots, \alpha_n]$ and determinant δ_n.

Proof. We proceed by induction on n. The theorem is trivial for $n = 1$. Let $n = 2$. Then elements ρ, σ of R may be determined so that $\rho\alpha_1 - \sigma\alpha_2 = \delta_2$, and thus we may choose

$$D_2 = \begin{bmatrix} \alpha_1 & \alpha_2 \\ \sigma & \rho \end{bmatrix}$$

Now suppose the theorem true for $n - 1, n \geq 3$, and let D_{n-1} be a matrix of R_{n-1} with first row $[\alpha_1, \alpha_2, \ldots, \alpha_{n-1}]$ and determinant $\delta_{n-1} = (\alpha_1, \alpha_2, \ldots, \alpha_{n-1})$. Since

$$\begin{aligned}\delta_n &= (\alpha_1, \alpha_2, \ldots, \alpha_n) \\ &= ((\alpha_1, \alpha_2, \ldots, \alpha_{n-1}), \alpha_n) \\ &= (\delta_{n-1}, \alpha_n)\end{aligned}$$

we can find elements ρ, σ of R such that $\rho\delta_{n-1} - \sigma\alpha_n = \delta_n$. Put

$$D_n = \begin{bmatrix} & & & & \alpha_n \\ & D_{n-1} & & & 0 \\ & & & & \cdot \\ & & & & \cdot \\ & & & & \cdot \\ & & & & 0 \\ \dfrac{\alpha_1\sigma}{\delta_{n-1}} & \dfrac{\alpha_2\sigma}{\delta_{n-l}} & \cdots & \dfrac{\alpha_{n-1}\sigma}{\delta_{n-1}} & \rho \end{bmatrix}$$

Then certainly D_n belongs to R_n, and has $[\alpha_1, \alpha_2, \ldots, \alpha_n]$ as its first row. If we expand by minors of the last column of D_n, we get

$$\det(D_n) = \rho \det(D_{n-1}) + (-1)^{n-1}\alpha_n \det(E_{n-1})$$

where E_{n-1} is the minor matrix of D_n obtained by striking out its first row and last column. Because of the nature of the first and last rows of D_n, we have

$$\delta_{n-1}E_{n-1} = \begin{bmatrix} 0 & \delta_{n-1} & 0 & \cdots & 0 \\ 0 & 0 & \delta_{n-1} & \cdots & 0 \\ \cdot & \cdot & \cdot & & \cdot \\ \cdot & \cdot & \cdot & & \cdot \\ \cdot & \cdot & \cdot & & \cdot \\ 0 & 0 & 0 & \cdots & \delta_{n-1} \\ \sigma & 0 & 0 & \cdots & 0 \end{bmatrix} D_{n-1}$$

so that

$$\begin{aligned}\delta_{n-1}^{n-1} \det(E_{n-1}) &= (-1)^{n-2}\sigma\, \delta_{n-1}^{n-2}\det(D_{n-1}) \\ &= (-1)^{n-2}\sigma\delta_{n-1}^{n-1}, \\ \det(E_{n-1}) &= (-1)^{n-2}\sigma\end{aligned}$$

Thus

$$\det(D_n) = \rho\delta_{n-1} - \sigma\alpha_n = \delta_n$$

Hence the theorem is proved for n, and the induction is complete.

This theorem has the useful corollary:

Corollary II.1. Suppose that $\alpha_1, \alpha_2, \ldots, \alpha_n$ are relatively prime elements of R. Then there is a unit matrix of R_n having $[\alpha_1, \alpha_2, \ldots, \alpha_n]$ as any prescribed row.

Of course the same result holds if a column is prescribed, since U belongs to R_n' if and only if U^T belongs to R_n'.

6. The Hermite normal form

We are now in a position to prove the fundamental result on one-sided equivalence, which we do for left equivalence only. Again, the situation for right equivalence is entirely similar.

Theorem II.2 (Hermite normal form). Every matrix A of R_n is left equivalent to a lower triangular matrix with each diagonal element lying in a prescribed complete set of nonassociates and with each element below the main diagonal lying in a prescribed complete set of residues modulo the diagonal element above it.

Proof. Let A be any element of R_n. We first show by induction on n that A is left equivalent to a lower triangular matrix. This is certainly true for $n = 1$. Assume then that any element of R_{n-1}, $n \geq 2$, is left equivalent to a lower triangular matrix.

Consider the last column of A. Either every element is zero, or there is a nonzero element. Suppose the latter. Let $[\alpha_1, \alpha_2, \ldots, \alpha_n]^T$ be the last column, and put $\delta = (\alpha_1, \alpha_2, \ldots, \alpha_n)$. Then certainly $\delta \neq 0$. Now elements $\rho_1, \rho_2, \ldots, \rho_n$ of R may be determined so that $\rho_1\alpha_1 + \rho_2\alpha_2 + \cdots + \rho_n\alpha_n = \delta$. Furthermore, $(\rho_1, \rho_2, \ldots, \rho_n) = 1$. By Corollary II.1, there is a unit matrix U_1 of R_n with last row $[\rho_1, \rho_2, \ldots, \rho_n]$. It follows that the (n, n) element of $A_1 = U_1A$ is δ, and that the remaining elements of the last column of A_1 are all divisible by δ. Hence by elementary operations of type (6), which correspond to left multiplication by some element U_2 of R_n', A_1 may be brought to the form

$$A_2 = U_2A_1 = \begin{bmatrix} & & & & 0 \\ & B & & & 0 \\ & & & & \cdot \\ & & & & \cdot \\ & & & & \cdot \\ & & & & 0 \\ * & * & \cdots & * & \delta \end{bmatrix}$$

where B belongs to R_{n-1}. By the induction hypothesis, there is an element V of R'_{n-1} such that VB is in lower triangular form. Then $U_3 = V \dotplus I_1$ belongs to R_n', and $A_3 = U_3 A_2$ is in lower triangular form. Thus the result has been proved for n, and the induction is complete. Hence A is left equivalent to a lower triangular matrix T.

In order to make the diagonal elements of T lie in any prescribed set of nonassociates, it is only necessary to multiply T on the left by

$$D = \operatorname{diag}(\varepsilon_1, \varepsilon_2, \ldots, \varepsilon_n)$$

for appropriate units $\varepsilon_1, \varepsilon_2, \ldots, \varepsilon_n$ of R; and this matrix belongs to R_n'.

In order to make each element of DT below the main diagonal lie in a prescribed complete set of residues modulo the diagonal element which lies about it, we first reduce the $(n, n-1)$ element modulo the $(n-1, n-1)$ element by adding a multiple of row $n-1$ to row n, then reduce the $(n, n-2)$ and $(n-1, n-2)$ elements modulo the $(n-2, n-2)$ element by adding multiples of row $n-2$ to rows n and $n-1$, etc., using elementary operations of type (6). This completes the proof.

Theorem II.2 is concerned with square matrices. However an examination of the proof will show that, with obvious modifications, the theorem can be stated and proved for any $(n+t) \times n$ matrix over R. In this case, the normal form can be taken as $\left[\begin{smallmatrix} H \\ 0 \end{smallmatrix}\right]$, where 0 is a $t \times n$ block of zeros and the $n \times n$ matrix H is in the form prescribed by Theorem II.2.

It is quite worthwhile to do a complete example. We choose $R = Z$. As a complete set of nonassociates for Z we choose the nonnegative integers, and as a complete set of residues modulo any nonzero integer m we choose

$$0, 1, \ldots, |m| - 1$$

Furthermore, a complete set of residues modulo 0 will be all of Z. Choose

$$A = \begin{bmatrix} 3 & 5 & 2 \\ -7 & 2 & 6 \\ 1 & 1 & 4 \end{bmatrix}$$

so that A belongs to Z_3. The matrices of Z_3' which arise as left multipliers in the reduction of A are

$$U_1 = \begin{bmatrix} 0 & 0 & 1 \\ 0 & 1 & 0 \\ 1 & 0 & 0 \end{bmatrix}, \qquad U_2 = \begin{bmatrix} 1 & 0 & -2 \\ 0 & 1 & -3 \\ 0 & 0 & 1 \end{bmatrix},$$

$$U_3 = \begin{bmatrix} -1 & 1 & 0 \\ -3 & 2 & 0 \\ 0 & 0 & 1 \end{bmatrix}, \qquad U_4 = \begin{bmatrix} 1 & 4 & 0 \\ 0 & 1 & 0 \\ 0 & 0 & 1 \end{bmatrix},$$

$$U_5 = \begin{bmatrix} -1 & 0 & 0 \\ 0 & 1 & 0 \\ 0 & 0 & 1 \end{bmatrix}, \qquad U_6 = \begin{bmatrix} 1 & 0 & 0 \\ 0 & 1 & 0 \\ 0 & -5 & 1 \end{bmatrix},$$

$$U_7 = \begin{bmatrix} 1 & 0 & 0 \\ 1 & 1 & 0 \\ -1 & 0 & 1 \end{bmatrix}$$

The steps in the reduction are

$$A_1 = U_1A = \begin{bmatrix} 1 & 1 & 4 \\ -7 & 2 & 6 \\ 3 & 5 & 2 \end{bmatrix}, \qquad A_2 = U_2A_1 = \begin{bmatrix} -5 & -9 & 0 \\ -16 & -13 & 0 \\ 3 & 5 & 2 \end{bmatrix},$$

$$A_3 = U_3A_2 = \begin{bmatrix} -11 & -4 & 0 \\ -17 & 1 & 0 \\ 3 & 5 & 2 \end{bmatrix}, \qquad A_4 = U_4A_3 = \begin{bmatrix} -79 & 0 & 0 \\ -17 & 1 & 0 \\ 3 & 5 & 2 \end{bmatrix},$$

$$A_5 = U_5A_4 = \begin{bmatrix} 79 & 0 & 0 \\ -17 & 1 & 0 \\ 3 & 5 & 2 \end{bmatrix}, \qquad A_6 = U_6A_5 = \begin{bmatrix} 79 & 0 & 0 \\ -17 & 1 & 0 \\ 88 & 0 & 2 \end{bmatrix},$$

$$A_7 = U_7A_6 = \begin{bmatrix} 79 & 0 & 0 \\ 62 & 1 & 0 \\ 9 & 0 & 2 \end{bmatrix}$$

Thus

$$\begin{bmatrix} 3 & 5 & 2 \\ -7 & 2 & 6 \\ 1 & 1 & 4 \end{bmatrix} \tilde{L} \begin{bmatrix} 79 & 0 & 0 \\ 62 & 1 & 0 \\ 9 & 0 & 2 \end{bmatrix}$$

The next theorem is concerned with uniqueness.

Theorem II.3. If A is a nonsingular element of R_n, then the Hermite normal form for A described by Theorem II.2 is unique.

Proof. Suppose that $A \tilde{L} H$, $A \tilde{L} K$, and that $H = (h_{ij})$, $K = (k_{ij})$ satisfy the hypotheses of Theorem II.2. Then there is a unit matrix $U = (u_{ij})$ such that $H = UK$. Multiply on the right by K^{adj} to obtain $\det(K)\cdot U = HK^{\text{adj}}$. Since H, K are nonsingular and lower triangular, U must also be lower triangular. Since $\det(U) = u_{11}u_{22}\ldots u_{nn}$ is a unit of R, each u_{ii} is a unit of R. But $h_{ii} = u_{ii}k_{ii}$, $1 \leq i \leq n$, and the elements h_{ii}, k_{ii} both belong to the same complete set of nonassociates. Thus each diagonal element of U must be 1, and so $h_{ii} = k_{ii}$, $1 \leq i \leq n$.

Suppose that $2 \leq s \leq n$. Then

$$h_{s,s-1} = \sum_{t=1}^{n} u_{st}k_{t,s-1} = \sum_{t=s-1}^{s} u_{st}k_{t,s-1}$$

since both U and K are lower triangular. Thus

$$\begin{aligned} h_{s,s-1} &= u_{s,s-1}k_{s-1,s-1} + u_{ss}k_{s,s-1} \\ &= u_{s,s-1}k_{s-1,s-1} + k_{s,s-1} \end{aligned}$$

It follows that

$$h_{s,s-1} \equiv k_{s,s-1} \bmod k_{s-1,s-1}$$

and so $h_{s,s-1} = k_{s,s-1}$, since they both belong to the same complete set of residues modulo $k_{s-1,s-1}$. Hence $u_{s,s-1} = 0$. We now proceed by induction. Suppose that we have shown that $u_{s,s-p} = 0$, $1 \leq p \leq j-1$. Then

$$\begin{aligned} h_{s,s-j} &= \sum_{t=1}^{n} u_{st}k_{t,s-j} = \sum_{t=s-j}^{s} u_{st}k_{t,s-j} \\ &= u_{s,s-j}k_{s-j,s-j} + u_{ss}k_{s,s-j} \end{aligned}$$

because both U and K are lower triangular and because of the induction hypothesis. Since $u_{ss} = 1$, we get

$$h_{s,s-j} \equiv k_{s,s-j} \bmod k_{s-j,s-j}$$

so that $h_{s,s-j} = k_{s,s-j}$, since they both belong to the same complete set of residues modulo $k_{s-j,s-j}$. Hence $u_{s,s-j} = 0$. Thus we have shown by induction that the off-diagonal elements of U are all 0. It follows that $U = I$, and $H = K$. This completes the proof.

7. Divisor sums

We will be concerned with divisor sums in the theorem that follows. Let f take on real or complex values, and be defined over R. The notation

$$\sum_{\delta|\mu} f(\delta)$$

will mean that δ runs over a complete set of nonassociated divisors of the nonzero element μ of R, all of which lie in some given complete set of nonassociates. In general f will be a *class function*, that is, will have the same value for associated arguments, and so the value of the sum will be independent of the particular complete set of nonassociates in which the divisors lie.

The function f will be said to be *arithmetically multiplicative* if it is a class function, and if $f(\mu\nu) = f(\mu)f(\nu)$ for all relatively prime elements μ, ν of R. If we choose our complete set of nonassociates to be closed with respect to multiplication (as we may be virtue of Corollary I.1) we can show that $F(\mu) = \sum_{\delta|\mu} f(\delta)$ is also such a function, since

$$\begin{aligned} F(\mu\nu) &= \sum_{\delta|\mu\nu} f(\delta) = \sum_{\delta_1|\mu,\delta_2|\nu} f(\delta_1\delta_2) \\ &= \sum_{\delta_1|\mu,\delta_2|\nu} f(\delta_1)f(\delta_2) = F(\mu)F(\nu) \end{aligned}$$

Furthermore, if $\mu = \pi^k$, π a prime, then

$$F(\pi^k) = f(1) + f(\pi) + \cdots + f(\pi^k)$$

These considerations apply when f is fully multiplicative as well, in which case the formula above simplifies to

$$F(\pi^k) = \frac{f(\pi)^{1+k} - 1}{f(\pi) - 1}$$

8. The Hermite normal form class number

In the theorem that follows we assume that R is a principal ideal ring such that $N(\mu)$ is finite for every nonzero element μ of R. We choose a complete set of nonassociates M for R which is multiplicatively closed, as in the previous discussion. If Δ is a fixed nonzero element of R, we define $H_n(\Delta)$ as the number of classes with respect to left equivalence of elements of R_n of determinant $\varepsilon\Delta$, where ε belongs to R' and is chosen so that $\varepsilon\Delta$ belongs to M.

Theorem II.4. Let Δ be a nonzero element of R. Then $H_n(\Delta)$ is a class function, and satisfies

$$H_1(\Delta) = 1, \tag{7}$$

$$H_n(\Delta) = \sum_{\delta|\Delta} N(\delta)H_{n-1}(\delta), \qquad n \geq 2 \tag{8}$$

It follows that $H_n(\Delta)$ is arithmetically multiplicative, and that if $\Delta = \pi^k$, π a prime, then

$$H_n(\pi^k) = \prod_{j=1}^{n-1} \frac{N(\pi)^{k+j} - 1}{N(\pi)^j - 1} \tag{9}$$

Proof. It is clear from the definition of $H_n(\Delta)$ that it is a class function and that $H_1(\Delta) = 1$. We may assume that Δ belongs to M. Then $H_n(\Delta)$ is just the number of matrices

$$\begin{bmatrix} h_1 & 0 & 0 & \cdots & 0 \\ h_{21} & h_2 & 0 & \cdots & 0 \\ h_{31} & h_{32} & h_3 & \cdots & 0 \\ \cdot & \cdot & \cdot & & \cdot \\ \cdot & \cdot & \cdot & & \cdot \\ \cdot & \cdot & \cdot & & \cdot \\ h_{n1} & h_{n2} & h_{n3} & \cdots & h_n \end{bmatrix}$$

such that h_j belongs to M, $1 \leq j \leq n$, h_{ij} lies in some specified complete set of residues modulo h_j for $j+1 \leq i \leq n$, and $h_1 h_2 \cdots h_n = \Delta$. Thus each h_{ij} can assume $N(h_j)$ values, and it follows that

$$H_n(\Delta) = \sum_{h_1 h_2 \cdots h_n = \Delta} N(h_1)^{n-1} N(h_2)^{n-2} \cdots N(h_{n-1})$$

If we rewrite this sum as

$$H_n(\Delta) = \sum_{h_1 h_2 \cdots h_n = \Delta} N(h_1 h_2 \cdots h_{n-1}) \cdot N(h_1)^{n-2} N(h_2)^{n-3} \cdots N(h_{n-2})$$

using the multiplicativity of the norm proved in Chapter I and note that

$$\sum_{h_1 h_2 \cdots h_n = \Delta} = \sum_{h_n|\Delta} \sum_{h_1 h_2 \cdots h_{n-1} = \Delta/h_n}$$

we get

$$H_n(\Delta) = \sum_{\delta|\Delta} N\left(\frac{\Delta}{\delta}\right) H_{n-1}\left(\frac{\Delta}{\delta}\right) = \sum_{\delta|\Delta} N(\delta) H_{n-1}(\delta)$$

This proves (8). Now choose $\Delta = \pi^k$, π a prime. Then (8) becomes

$$H_n(\pi^k) = \sum_{j=0}^{k} N(\pi)^j H_{n-1}(\pi^j)$$

This implies that

$$H_n(\pi^k) = N(\pi^k)H_{n-1}(\pi^k) + H_n(\pi^{k-1}) \tag{10}$$

Together with

$$H_1(\pi^k) = H_n(1) = 1 \tag{11}$$

this easily yields (9) by double induction on n and k. Finally, (8) implies by induction on n that $H_n(\Delta)$ is arithmetically multiplicative. This completes the proof.

9. An application of the Hermite normal form

An interesting application of Theorem II.2 is to show that every left ideal of R_n is principal for any principal ideal ring R. We require a lemma:

Lemma II.1. Let R be a principal ideal ring. Let V be an R-module which is finitely generated over R. Then every submodule of V is also finitely generated over R.

Proof. Let $v_1, v_2, \ldots, v_n$ be generators for V over R. Let W be any submodule of V. For each k such that $1 \leq k \leq n$, let W_k be the set of elements of W of the form

$$w = \alpha_k v_k + \alpha_{k+1} v_{k+1} + \cdots + \alpha_n v_n$$

where $\alpha_k, \alpha_{k+1}, \ldots, \alpha_n$ belong to R. Either $\alpha_k = 0$ for every such element w of W_k, or there is at least one for which $\alpha_k \neq 0$. In the latter case, choose one for which the number of prime divisors of α_k is least; say

$$w_k = \alpha_{kk} v_k + \alpha_{k,k+1} v_{k+1} + \cdots + \alpha_{kn} v_n$$

In the former case, define w_k to be 0. Suppose now that we are in the latter case. Then α_{kk} divides α_k for every element w of W_k. To see this, put $(\alpha_k, \alpha_{kk}) = \delta$, and determine elements ρ, σ of R such that $\rho\alpha_k + \sigma\alpha_{kk} = \delta$. Then certainly $\delta \neq 0$, and

$$\rho w + \sigma w_k = \delta v_k + \cdots$$

belongs to W_k. But if α_{kk} did not divide α_k, δ would have fewer prime divisors than α_{kk}, which would contradict the definition of α_{kk}. Thus α_{kk} does divide α_k for every element w of W_k. It follows that the elements $w_1, w_2, \ldots, w_n$ generate W. For if w is any element of W, then we may determine β_1 in R such that $w - \beta_1 w_1$ belongs to W_2; and then determine β_2 in R such that $w - \beta_1 w_1 - \beta_2 w_2$ belongs to W_3; etc. Thus $w = \beta_1 w_1 + \beta_2 w_2 + \cdots + \beta_n w_n$ and so $w_1, w_2, \ldots, w_n$ generate W.

We now prove

Theorem II.5. Let R be a principal ideal ring. Then every left ideal of R_n is principal.

Proof. Considered as an R-module, R_n has the finite R-basis E_{ij}, $1 \leq i, j \leq n$, where E_{ij} is the matrix with 1 in position (i, j) and 0 elsewhere. Let A be a left ideal of R_n. Then A is a submodule of R_n, and so by Lemma II.1 A is generated by finitely many elements A_1, $A_2, \ldots, A_t$ of A, where $t \leq n^2$. Consider the $nt \times n$ matrix

$$M = \begin{bmatrix} A_1 \\ A_2 \\ \cdot \\ \cdot \\ \cdot \\ A_t \end{bmatrix}$$

Let P be an element of R_{nt}' such that PM is in Hermite norm form. Write $P = (P_{ij})$, where each P_{ij} belongs to R_n. Then

$$PM = \begin{bmatrix} H \\ 0 \end{bmatrix}$$

where the zero block is $(nt - n) \times n$ and H is an $n \times n$ matrix in Hermite normal form. Since $P_{11}A_1 + P_{12}A_2 + \cdots + P_{1t}A_t = H$, H belongs to A. Hence $A \supset (H)$. Set $P^{-1} = Q = (Q_{ij})$. Then Q also belongs to R_{nt}', and each Q_{ij} belongs to R_n. Then

$$Q\begin{bmatrix} H \\ 0 \end{bmatrix} = M$$

and so $Q_{i1}H = A_i$, $1 \leq i \leq t$. Hence $(H) \supset A$, and so $A = (H)$. This completes the proof.

10. Left equivalence over a euclidean ring

We now consider left equivalence and the Hermite normal form for the case when the underlying ring R is euclidean. It is here that the full importance of the elementary operations, or elementary matrices, becomes evident. In fact, we prove

Theorem II.6. Let R be euclidean. Then every element of R_n may be brought to Hermite normal form by elementary row operations alone.

Proof. We first show by induction on n that every element of R_n may be brought to lower triangular form by elementary row operations alone. This is certainly true for $n = 1$. Assume the truth of this statement for $n - 1$, $n \geq 2$. Let A be any element of R_n. Consider the last column of A. Either every element is zero, or there is a nonzero element. Suppose the latter. Choose a nonzero element with minimal v and put it into the (n, n) position by a suitable interchange of rows. Now subtract multiples of the last row from all the other rows in such a manner as to make the remaining elements of the last column 0, or have smaller v than that of the (n, n) element. After finitely many repetitions of this process, A must be brought to a form in which all the elements of the last column are 0, with the exception of the (n, n) element. By the induction hypothesis applied to the $(n - 1) \times (n - 1)$ matrix obtained by deleting the last row and column, it follows that A may be brought into lower triangular form by elementary row operations alone. Thus the induction is complete and we have shown that every element of R_n may be brought to lower triangular form by elementary row operations alone.

The rest of the proof is the same as in Theorem II.2, since all the operations involved there are elementary row operations. This completes the proof.

11. Generators of R_n' when R is euclidean

If we choose a complete set of nonassociates for R so that it contains 1, as we have agreed to do, and a complete set of residues modulo 1 to be 0, then the Hermite normal form of any element of R_n' is just I. It follows that any unimodular matrix of R_n may be brought to the form I by elementary row operations when R is euclidean. Since the inverses of the elementary row matrices are again elementary row matrices, we get

Theorem II.7. Let R be a euclidean ring. Then every unit matrix of R_n is the product of finitely many elementary row matrices; that is R_n' is generated by the elementary row matrices.

The problem of determining minimal sets of generators for R_n', R euclidean, is interesting and difficult. Since the permutation matrices may be generated by

$$T = \begin{bmatrix} 0 & 1 \\ 1 & 0 \end{bmatrix} \dotplus I_{n-2},$$

$$P = \begin{bmatrix} 0 & 1 & 0 & \cdots & 0 \\ 0 & 0 & 1 & \cdots & 0 \\ \cdot & \cdot & \cdot & & \cdot \\ \cdot & \cdot & \cdot & & \cdot \\ \cdot & \cdot & \cdot & & \cdot \\ 0 & 0 & 0 & \cdots & 1 \\ 1 & 0 & 0 & \cdots & 0 \end{bmatrix}$$

It suffices to take T, P, and the matrices

$$D(\varepsilon) = \operatorname{diag}(\varepsilon, 1, \ldots, 1),$$

$$S(\mu) = \begin{bmatrix} 1 & \mu \\ 0 & 1 \end{bmatrix} \dotplus I_{n-2}$$

where ε runs over R' and μ over R.

In many important instances R is finitely generated as a Z-module, and R' is a finitely generated group; in this case R_n' is finitely generated. Thus if $R = Z$, it suffices to take T, P, $D(-1)$, and $S(1)$; and Hua and Reiner have shown [19] that these may be reduced further to $D(-1)$, $S(1)$, and

$$P_1 = \begin{bmatrix} 0 & 1 & 0 & \cdots & 0 \\ 0 & 0 & 1 & \cdots & 0 \\ \cdot & \cdot & \cdot & & \cdot \\ \cdot & \cdot & \cdot & & \cdot \\ \cdot & \cdot & \cdot & & \cdot \\ 0 & 0 & 0 & \cdots & 1 \\ (-1)^{n-1} & 0 & 0 & \cdots & 0 \end{bmatrix}$$

In fact, they showed that the group of $n \times n$ matrices over Z of determinant 1 may be generated by $S(1)$, P_1.

12. Two-sided equivalence

We now turn to two-sided equivalence. For this purpose we require some elementary material about determinants, which we summarize, but do not prove.

Let S be any commutative ring. Let k be any integer such that $1 \leq k \leq n$. Let $Q_{k,n}$ be the set of all k-tuples $\{i_1, i_2, \ldots, i_k\}$, where $i_1, i_2, \ldots, i_k$ are integers such that $1 \leq i_1 < i_2 < \cdots < i_k \leq n$. Then $Q_{k,n}$ contains $\binom{n}{k}$ elements. Let A be any element of S_n. If ω, τ are elements of $Q_{k,n}$, then $A(\omega, \tau)$ will denote the $k \times k$ determinantal minor of A whose row indices are the elements of ω and whose column indices are the elements of τ. There are clearly $\binom{n}{k}^2$ such minors. The *Cauchy–Binet theorem* states that if $C = AB$, where A, B are elements of S_n, then

$$C(\omega, \tau) = \sum_{\alpha \in Q_{k,n}} A(\omega, \alpha)B(\alpha, \tau) \tag{12}$$

[which is entirely equivalent to the statement that $C_k(AB) = C_k(A)\, C_k(B)$, where $C_k(A)$ is the kth compound of A].

13. Determinantal divisors

We return to the principal ideal ring R. Let A be any element of R_n. Let k be any integer such that $1 \leq k \leq n$. If $A(\omega, \tau) = 0$ for all ω, τ of $Q_{k,n}$, we put $d_k(A) = 0$; if not, we put $d_k(A)$ equal to the greatest common divisor of the $\binom{n}{k}^2$ elements $A(\omega, \tau)$; $d_k(A)$ is called the kth *determinantal divisor* of A.

14. Multiples

Let A, B be elements of R_n. If there are elements P, Q of R_n such that $B = PAQ$, we say that B is a *multiple* of A.

We have

Lemma II.2. Suppose that B is a multiple of A. Then if $d_k(A) = 0$,

$d_k(B) = 0$. If $d_k(A) \neq 0$, then $d_k(A) \mid d_k(B)$. Here k is any integer such that $1 \leq k \leq n$.

Proof. We have that $B = PAQ$ for elements P, Q of R_n. By the Cauchy–Binet theorem,

$$B(\omega, \tau) = \sum_{\alpha, \beta \in Q_{k,n}} P(\omega, \alpha)\, A(\alpha, \beta)\, Q(\beta, \tau)$$

If $d_k(A) = 0$ then $A(\alpha, \beta) = 0$ for all α, β of $Q_{k,n}$, and so $B(\omega, \tau) = 0$ for all ω, τ of $Q_{k,n}$, so that $d_k(B) = 0$. If $d_k(A) \neq 0$, then $d_k(A) \mid A(\alpha, \beta)$ for all α, β of $Q_{k,n}$ and so $d_k(A) \mid B(\omega, \tau)$ for all ω, τ of $Q_{k,n}$. Hence $d_k(A) \mid d_k(B)$, and the proof is complete.

The importance of the determinantal divisors is evident from the following theorem:

Theorem II.8. Suppose that A, B belong to R_n, and that $A \tilde{E} B$. Then $d_k(A) = d_k(B)$, $1 \leq k \leq n$.

Proof. Since $A \tilde{E} B$, there exist unimodular matrices U, V of R_n such that $A = UBV$; and hence also that $B = U^{-1}AV^{-1}$. Thus A is a multiple of B and B is a multiple of A. It follows by the previous theorem that $d_k(A)$ and $d_k(B)$ are associates, $1 \leq k \leq n$. Since we have made the convention that greatest common divisors must always lie in some prescribed complete set of nonassociates, $d_k(A) = d_k(B)$, $1 \leq k \leq n$. This completes the proof.

15. The Smith normal form

We now come to the most important theorem in this area, and quite possibly the most important theorem in all of elementary matrix theory.

Theorem II.9 (Smith normal form). Every matrix A of R_n is equivalent to a diagonal matrix

$$S = S(A) = \operatorname{diag}(s_1, s_2, \ldots, s_r, 0, 0, \ldots, 0)$$

where r is the rank of A, $s_1, s_2, \ldots, s_r$ are nonzero elements of R (which may be taken to belong to any prescribed complete set of nonassociates), and $s_i \mid s_{i+1}$, $1 \leq i \leq r - 1$.

Proof. We first note that if the rank of a matrix of R_n is defined as the size of a largest nonvanishing determinantal minor, then the Cauchy–Binet theorem readily implies that equivalent matrices have the same rank.

We may assume that A contains a nonzero element, which may be brought to the (1, 1) position by suitable row and column interchanges. As in the proof of Theorem II.2, this element may be replaced by the greatest common divisor of the first column. If the new (1, 1) element does not divide all the elements of the first row, then it may be replaced by the greatest common divisor of the elements of the first row, and will contain fewer prime factors than before. The process is repeated until an element in the (1, 1) position is obtained which divides every element of the first row and column. By elementary row and column operations all the elements in the first row and column, other than the (1, 1) element, can be made zero. Denote this new matrix, which is equivalent to A, by $C = (c_{ij})$.

Suppose that the submatrix of C obtained by deleting the first row and column contains an element c_{ij} which is not divisible by c_{11}. Add column j to column 1. Column 1 then consists of the elements c_{11}, $c_{2j}, \ldots, c_{ij}, \ldots, c_{nj}$. Repeating the previous process we can replace c_{11} by a proper divisor of itself. Thus we must finally reach the stage where the element in the (1, 1) position divides every element of the matrix, and all the other elements of the first row and column are zero.

The entire process is now repeated with the submatrix obtained by deleting the first row and column. Eventually a stage is reached when the matrix has the form

$$\begin{bmatrix} D & 0 \\ 0 & E \end{bmatrix}$$

where $D = \operatorname{diag}(s_1, s_2, \ldots, s_r)$, and $s_i \mid s_{i+1}$, $1 \leq i \leq r - 1$. But then E must be the zero matrix, since otherwise A would have rank larger than r. This completes the proof.

Again, Theorem II.9 has an obvious extension to the case when the matrix is not square. When R is euclidean, the reduction to canonical form may be accomplished by elementary row and column operations.

It is clear that with the proper choice of unit multipliers,

$$\begin{aligned} s_1 &= d_1(S) = d_1(A), \\ s_1 s_2 &= d_2(S) = d_2(A), \\ &\;\;\vdots \\ s_1 s_2 \cdots s_r &= d_r(S) = d_r(A), \\ d_k(S) &= d_k(A) = 0, \qquad r+1 \le k \le n \end{aligned}$$

Hence

$$\begin{aligned} s_1 &= d_1(A), \\ s_2 &= d_2(A)/d_1(A), \\ &\;\;\vdots \\ s_r &= d_r(A)/d_{r-1}(A) \end{aligned}$$

This guarantees that the canonical form $S(A)$ described by Theorem II.9 is unique, and yields

Theorem II.10. Two matrices A, B of R_n are equivalent if and only if they have the same determinantal divisors.

16. Invariant factors

Incidentally, the discussion shows that $d_k(A)$ divides $d_{k+1}(A)$, $1 \le k \le r-1$, which is obvious from the definition. Put $d_0(A) = 1$, for convenience. The quantities

$$s_k(A) = \frac{d_k(A)}{d_{k-1}(A)} \tag{13}$$

are known as the *invariant factors* of A. A knowledge of the invariant factors of A thus presupposes a knowledge of the rank of A.

Since $s_k(A)$ divides $s_{k+1}(A)$, $1 \le k \le r-1$, we have also shown that

$$d_k^2(A) \mid d_{k-1}(A) d_{k+1}(A), \qquad 1 \le k \le r-1$$

Theorem II.10 may be restated as

Theorem II.11. Two matrices A, B of R_n are equivalent if and only if they have the same invariant factors, apart from unit multipliers of R.

A well-known corollary is that two matrices of the same size over a field are equivalent if and only if they have the same rank, and any matrix of rank r is equivalent to $I_r \dotplus 0$.

17. Elementary divisors

Let A be any element of R_n of rank r and $s_k = s_k(A)$, $1 \leq k \leq r$, be its invariant factors. Let $\pi_1, \pi_2, \ldots, \pi_t$ be a complete set of non-associated primes which occur as divisors of the invariant factors. Then for appropriate nonnegative integers e_{ij} and appropriate units ε_i of R, we have

$$\begin{aligned} s_1 &= \varepsilon_1 \pi_1^{e_{11}} \pi_2^{e_{12}} \cdots \pi_t^{e_{1t}}, \\ s_2 &= \varepsilon_2 \pi_1^{e_{21}} \pi_2^{e_{22}} \cdots \pi_t^{e_{2t}}, \\ &\ \ \vdots \\ s_r &= \varepsilon_r \pi_1^{e_{r1}} \pi_2^{e_{r2}} \cdots \pi_t^{e_{rt}} \end{aligned}$$

It is clear from the divisibility properties of the invariant factors that

$$(14) \qquad 0 \leq e_{1j} \leq e_{2j} \leq \cdots \leq e_{rj}, \qquad 1 \leq j \leq t$$

The set of prime powers $\pi_j^{e_{ij}}$, $1 \leq i \leq r$, $1 \leq j \leq t$, including repetitions but excluding those with zero exponents, is called the set of *elementary divisors* of A. If the exponents which occur in this set are all 1, then A is said to have *simple* elementary divisors.

Given the set of elementary divisors, we can reconstruct the invariant factors because of the ordering condition (14). Thus if

$$e_j = \max_{1 \leq i \leq r} e_{ij}, \qquad 1 \leq j \leq t$$

then s_r must be an associate of $\pi_1^{e_1} \pi_2^{e_2} \cdots \pi_t^{e_t}$. Deleting these prime powers from the set of elementary divisors, we determine s_{r-1} in similar fashion, and so on. We thus are led to the following version of the fundamental theorem:

Theorem II.12. Two elements A, B of R_n are equivalent if and only if they have the same rank and same elementary divisors, apart from unit multipliers of R.

For example, take $R = Z$ and suppose that the matrix A of Z_n has rank 5 and elementary divisors $\{2^2, 2^2, 2^3, 2^3, 3, 3^2, 3^2, 3^2, 5, 5, 7\}$. Then $s_5 = 2^3 \cdot 3^2 \cdot 5 \cdot 7$. Deleting $2^3, 3^2, 5, 7$, we have the set $\{2^2, 2^2, 2^3, 3, 3^2, 3^2, 5\}$. Then $s_4 = 2^3 \cdot 3^2 \cdot 5$. Deleting $2^3, 3^2, 5$, we have the set $\{2^2, 2^2, 3, 3^2\}$. Then $s_3 = 2^2 \cdot 3^2$. Deleting $2^2, 3^2$, we have the set $\{2^2, 3\}$. Hence $s_2 = 2^2 \cdot 3$ and since the set obtained by deleting $2^2, 3$ is empty, s_1 must be 1.

The next theorem is useful if an equivalent diagonal matrix is known.

Theorem II.13. Suppose that A belongs to R_n, is of rank r, and is equivalent to $\operatorname{diag}(\lambda_1, \lambda_2, \ldots, \lambda_r, 0, 0, \ldots, 0)$. Then the prime power factors of the λ_i, $1 \le i \le r$, are the elementary divisors of A.

Proof. Let π be any prime which divides some λ_i, $1 \le i \le r$. Arrange the λ_i according to ascending powers of π:

$$\begin{array}{ll} \lambda_{i_1} = \pi^{e_1}\mu_1, & (\pi, \mu_1) = 1 \\ \lambda_{i_2} = \pi^{e_2}\mu_2, & (\pi, \mu_2) = 1 \\ \vdots & \\ \lambda_{i_r} = \pi^{e_r}\mu_r, & (\pi, \mu_r) = 1 \end{array}$$

so that $0 \le e_1 \le e_2 \le \cdots \le e_r$. Then clearly, the exact power of π which divides $d_k(A)$ is $\pi^{e_1+e_2+\cdots+e_k}$, $1 \le k \le r$. Hence the exact power of π which divides $s_k(A) = d_k(A)/d_{k-1}(A)$ is π^{e_k}, $1 \le k \le r$ [recall that $d_0(A) = 1$]. Thus π^{e_k} is an elementary divisor, $1 \le k \le r$.

Applying this argument for all primes π which divide some λ_i, we obtain the result.

As an example, choose $R = F[x]$ and suppose that A belongs to R_6, is of rank 4, and is equivalent to

$$\operatorname{diag}(x(x-1)^2(x+1)^2, (x-1)^2(x+1)^2, x^2(x-1), x(x+1)^3, 0, 0)$$

Then the elementary divisors of A are

$$x, x, x^2, x-1, (x-1)^2, (x-1)^2, (x+1)^2, (x+1)^2, (x+1)^3$$

Thus the invariant factors of A are

$$s_4 = x^2(x-1)^2(x+1)^3, \qquad s_3 = x(x-1)^2(x+1)^2,$$
$$s_2 = x(x-1)(x+1)^2, \qquad s_1 = 1$$

Hence

$$A \mathrel{\tilde{E}} \operatorname{diag}(1, x(x-1)(x+1)^2, x(x-1)^2(x+1)^2, x^2(x-1)^2(x+1)^3, 0, 0)$$

18. Divisibility properties of the invariant factors

Let A, B be nonsingular elements of R_n. A consequence of Lemma II.2 is that $d_k(AB)$ is divisible by $d_k(A)$ and by $d_k(B)$, $1 \le k \le n$. The same result holds for the invariant factors: $s_k(AB)$ is divisible by $s_k(A)$ and by $s_k(B)$, $1 \le k \le n$. We will prove this below, as well as the interesting consequence that $S(AB) = S(A)S(B)$ provided A and B have relatively prime determinants, and $S(A)$ is the normal form specified by Theorem II.9 in the case when A is nonsingular.

We also note that the result concerning the determinantal divisors follows as a corollary, since if A is nonsingular then its kth determinantal divisor $d_k(A)$ is just $s_1(A)s_2(A) \cdots s_k(A)$, $1 \le k \le n$.

In what follows congruence of matrices will mean elementwise congruence. We first prove the following lemma:

Lemma II.3. Suppose that $\left[\begin{smallmatrix} H & 0 \\ 0 & K \end{smallmatrix}\right]$ is a nonsingular element of R_n which is in Smith normal form. Let m be any nonzero element of R and suppose that there is a matrix U of R_n such that $(\det(U), m) = 1$, and

$$U\begin{bmatrix} H & 0 \\ 0 & K \end{bmatrix} \equiv \begin{bmatrix} * & * \\ 0 & 0 \end{bmatrix} \bmod m \tag{15}$$

Then $K \equiv 0 \bmod m$.

Proof. Put

$$H = \operatorname{diag}(h_1, h_2, \ldots, h_r), \qquad K = \operatorname{diag}(k_1, k_2, \ldots, k_s)$$

where $r + s = n$ and $h_i \mid h_{i+1}$ $(1 \le i \le r-1)$, $k_j \mid k_{j+1}$ $(1 \le j \le s-1)$, $h_r \mid k_1$. Partition U so that

$$U = \begin{bmatrix} U_1 & U_2 \\ U_3 & U_4 \end{bmatrix}$$

where U_1 is $r \times r$, and U_4 is $s \times s$. Then (15) implies that

(16) $$U_3 H \equiv 0 \bmod m,$$

(17) $$U_4 K \equiv 0 \bmod m$$

We can multiply (16) on the right by $\operatorname{diag}(h_r/h_1, h_r/h_2, \ldots, 1)$ to obtain

$$h_r U_3 \equiv 0 \bmod m$$

Put

(18) $$(h_r, m) = \delta$$

Then

$$(h_r/\delta) U_3 \equiv 0 \bmod m/\delta$$

and since $(h_r/\delta, m/\delta) = 1$,

(19) $$U_3 \equiv 0 \bmod m/\delta$$

Now set $K = k_1 K'$, where $K' = \operatorname{diag}(1, k_2/k_1, \ldots, k_s/k_1)$. Then from (17),

$$k_1 U_4 K' \equiv 0 \bmod m$$

Put

(20) $$(k_1, m) = \Delta$$

Then as before we deduce that

(21) $$U_4 K' \equiv 0 \bmod m/\Delta$$

Now $\delta \mid \Delta$, in virtue of (18), (20), and the fact that $h_r \mid k_1$. It follows that $m/\Delta \mid m/\delta$, and so (19) holds modulo m/Δ as well. Thus

$$U = \begin{bmatrix} U_1 & U_2 \\ U_3 & U_4 \end{bmatrix} \equiv \begin{bmatrix} U_1 & U_2 \\ 0 & U_4 \end{bmatrix} \bmod \frac{m}{\Delta},$$

$$\det(U) \equiv \det(U_1) \det(U_4) \bmod m/\Delta$$

Since $(\det(U), m) = 1$, it follows that $(\det(U_4), m/\Delta) = 1$. Hence (21) implies that $K' \equiv 0 \bmod m/\Delta$; and so $\Delta \equiv 0 \bmod m$, since the (1, 1) element of K' is 1. Thus (20) implies that $m \mid k_1$ and the conclusion follows.

We are now prepared to use the lemma. Let A, B be nonsingular elements of R_n. Then matrices U, C of R_n exist such that U is a unit matrix and

(22) $$US(AB) = S(A)C$$

Certainly $s_1(A)$ divides $s_1(AB)$, since $s_1(A)$ is the greatest common divisor of the elements of A and $s_1(AB)$ the greatest common divisor of the elements of AB. For $2 \leq k \leq n-1$, choose $m = s_k(A)$ and apply Lemma II.3. We are left with $k = n$. Write $U = (u_{ij})$, $C = (c_{ij})$. Then (22) implies that

$$u_{ij}s_j(AB) = c_{ij}s_i(A)$$

so that for $i = n$,

$$u_{nj}s_j(AB) \equiv 0 \bmod s_n(A)$$

Since $s_j(AB) \mid s_n(AB)$, $1 \leq j \leq n$, this implies that

$$u_{nj}s_n(AB) \equiv 0 \bmod s_n(A)$$

The fact that $s_n(A) \mid s_n(AB)$ now follows, since $(u_{n1}, u_{n2}, \ldots, u_{nn}) = 1$.

If we note that in addition $S(A^T) = S(A)$, the entire argument may also be applied to the pair B^T, A^T and we finally obtain

Theorem II.14. Let A, B be nonsingular elements of R_n. Then $s_k(AB)$ is divisible by $s_k(A)$ and by $s_k(B)$ for $1 \leq k \leq n$.

19. The multiplicativity of the Smith normal form

From this theorem we easily deduce

Theorem II.15. Suppose that A, B are elements of R_n with relatively prime determinants. Then

$$S(AB) = S(A)S(B)$$

Proof. Since $(\det(A), \det(B)) = 1$ and $s_k(A) \mid \det(A)$, $s_k(B) \mid \det(B)$, it follows that $(s_k(A), s_k(B)) = 1$ for $1 \leq k \leq n$. Then the previous theorem implies that

$$s_k(AB) \equiv 0 \bmod s_k(A)s_k(B), \qquad 1 \leq k \leq n$$

But $\prod_{k=1}^n s_k(AB)$ is a unit multiple of $\det(AB)$, and $\prod_{k=1}^n s_k(A)s_k(B)$ is a unit multiple of $\det(A)\det(B)$. It follows that $s_k(AB)/s_k(A)s_k(B)$ is a unit for $1 \leq k \leq n$. But this implies that in fact $s_k(AB) = s_k(A)s_k(B)$ for $1 \leq k \leq n$, since associated elements in corresponding diagonal positions of matrices in Smith normal form must be equal. This completes the proof.

The theorem is definitely false if $(\det(A), \det(B)) > 1$. Thus if m is any element of R, and

$$A = \begin{bmatrix} 1 & 1 \\ 0 & m \end{bmatrix}, \qquad B = \begin{bmatrix} 1 & 0 \\ -1 & m \end{bmatrix}$$

then

$$S(A) = S(B) = \begin{bmatrix} 1 & 0 \\ 0 & m \end{bmatrix};$$

but

$$S(AB) = \begin{bmatrix} m & 0 \\ 0 & m \end{bmatrix}$$

This example also shows that $S(AB)$ need not equal $S(BA)$, since here

$$S(BA) = \begin{bmatrix} 1 & 0 \\ 0 & m^2 \end{bmatrix}$$

However, $S(AB)$ and $S(BA)$ are equal if A and B have relatively prime determinants as is readily seen from the preceding theorem.

20. The Smith normal form class number

We are now going to count the number $S_n(\Delta)$ of two-sided equivalence classes of matrices of R_n of fixed nonzero determinant Δ, which is clearly a class function. Before doing so, however, we make some remarks of an elementary number-theoretic nature.

Let n be any positive integer. The number of partitions of n into k nonnegative summands $x_1, x_2, \ldots, x_k$ such that $x_1 \leq x_2 \leq \cdots \leq x_k$ will be denoted by $p(n; k)$. Thus

$$p(n; k) = \sum_{\substack{x_1+x_2+\cdots+x_k=n \\ 0\leq x_1\leq x_2\leq\cdots\leq x_k}} 1$$

Put

$$\begin{aligned} x_1 &= y_1, \\ x_2 &= y_1 + y_2, \\ &\vdots \\ x_k &= y_1 + y_2 + \cdots + y_k \end{aligned}$$

Then

$$p(n; k) = \sum_{\substack{ky_1+(k-1)y_2+\cdots+y_k=n \\ y_1\geq 0, y_2\geq 0,\ldots, y_k\geq 0}} 1$$

In this form, it is at once obvious that $p(n; k)$ is the coefficient of x^n in

$$\{(1 - x)(1 - x)^2 \cdots (1 - x^k)\}^{-1}$$

so that $p(n; k)$ is just the number of partitions of n into parts not exceeding k, or equivalently, into at most k parts.

We return to the determination of $S_n(\Delta)$. Let

$$\Delta = \varepsilon \pi_1^{e_1} \pi_2^{e_2} \cdots \pi_t^{e_t}$$

be the prime power decomposition of Δ, where ε is a unit of R. Then $S_n(\Delta)$ is just the number of ways of forming

$$s_1(A) = \varepsilon_1 \pi_1^{e_{11}} \pi_2^{e_{12}} \cdots \pi_t^{e_{1t}},$$
$$s_2(A) = \varepsilon_2 \pi_1^{e_{21}} \pi_2^{e_{22}} \cdots \pi_t^{e_{2t}},$$
$$\vdots$$
$$s_n(A) = \varepsilon_n \pi_1^{e_{n1}} \pi_2^{e_{n2}} \cdots \pi_t^{e_{nt}}$$

where $\varepsilon_1, \varepsilon_2, \ldots, \varepsilon_n$ are units of R and the nonnegative integers e_{ij} satisfy

$$(23) \quad 0 \le e_{1j} \le e_{2j} \le \cdots \le e_{nj}, \qquad e_{1j} + e_{2j} + \cdots + e_{nj} = e_j$$

For each j such that $1 \le j \le t$, (23) has $p(e_j; n)$ solutions, by the preceding discussion; and hence the total number of ways of forming the invariant factors is

$$S_n(\Delta) = \prod_{j=1}^{t} p(e_j; n),$$

since the disposition of the units is irrelevant. Thus we have proved

Theorem II.16. Let $S_n(\Delta)$ denote the number of two-sided equivalence classes of matrices of R_n of fixed nonzero determinant Δ. Then $S_n(\Delta)$ is an arithmetically multiplicative function of Δ whose value at a prime power π^k is $p(k; n)$, the number of partitions of k into at most n parts.

21. Applications of the Smith normal form

We now go on to some applications of the Smith normal form, drawn from various areas.

(a) Let G be a finitely generated abelian group, H a subgroup of G. Then H is also finitely generated, and the rank of H does not exceed the rank of G. Let $x_1, x_2, \ldots, x_n$ be a basis for G, and $y_1, y_2, \ldots, y_m$, $m \leq n$, a basis for H. Then for some appropriate rational integral $m \times n$ matrix A,

$$y = Ax$$

where

$$x = [x_1, x_2, \ldots, x_n]^{\mathrm{T}}, \qquad y = [y_1, y_2, \ldots, y_m]^{\mathrm{T}}$$

We may determine rational integral unimodular matrices U, V such that U is $m \times m$, V is $n \times n$, and $A = UDV$, where

$$D = \operatorname{diag}(d_1, d_2, \ldots, d_m)$$

and $d_k \mid d_{k+1}$, $1 \leq k \leq m-1$. Thus $y = UDVx$. Put

$$u = U^{-1}y = [u_1, u_2, \ldots, u_m]^{\mathrm{T}}, \qquad v = VX = [v_1, v_2, \ldots, v_n]^{\mathrm{T}}$$

Then $v_1, v_2, \ldots, v_n$ is also a basis for G, $u_1, u_2, \ldots, u_m$ is also a basis for H, and

$$\begin{aligned} u_1 &= d_1 v_1, \\ u_2 &= d_2 v_2, \\ &\;\;\vdots \\ u_m &= d_m v_m \end{aligned}$$

That is, if H is any given subgroup of G then it is always possible to find a basis $v_1, v_2, \ldots, v_n$ of G such that $d_1 v_1, d_2 v_2, \ldots, d_m v_m$ is a basis for H.

(b) Let m be a given nonzero rational integer, and A a matrix of Z_n such that $\det(A) \equiv 1 \bmod m$. We will show that a matrix B of Z_n may be found such that $B \equiv A \bmod m$, and $\det(B) = 1$.

Put $A = UDV$, where U, V are unit matrices of Z_n such that $\det(UV) = 1$, and D is diagonal (we do not need the full Smith norm form). Then it is sufficient to show that $C \in Z_n$ can be determined such that $C \equiv D \bmod m$ and $\det(C) = 1$; for then $B = UCV \equiv UDV \equiv A \bmod m$ and $\det(B) = 1$. For this purpose, suppose that $D = \operatorname{diag}(d_1,$

$d_2, \ldots, d_n)$. Then $d_1 d_2 \cdots d_n = 1 + md$, say. Put

$$C = \begin{bmatrix} d_1 + mx & m & & \\ & d_2 & m & \\ & \cdots & & \\ my & & & d_n \end{bmatrix}$$

Then certainly $C \equiv D \bmod m$ and

$$\begin{aligned} \det(C) &= (d_1 + mx)d_2 \cdots d_n + m^n y \\ &= 1 + m\{d + d_2 \cdots d_n x + m^{n-1} y\} \end{aligned}$$

Since $(d_2 \cdots d_n, m) = 1$, integers x and y can be determined so that $d + d_2 \cdots d_n x + m^{n-1} y = 0$. For this choice of x and y, $\det(C) = 1$.

The same result holds, with the same proof, for any principal ideal ring R.

(c) An interesting application of the Smith normal form is to diophantine analysis. Let A be an $r \times n$ matrix over Z, c an $r \times 1$ vector over Z. Consider the problem of finding all $n \times 1$ vectors x over Z which satisfy

(24) $$Ax = c$$

We assume that $r \leq n$ and that A is of rank r. Then there are unimodular matrices U, V over Z such that U is $r \times r$, V is $n \times n$, and $UAV = [S \ \ 0]$ is in Smith normal form, where $S = \operatorname{diag}(s_1, s_2, \ldots, s_r)$, $s_i \mid s_{i+1}$, $1 \leq i \leq r - 1$. Set $V^{-1}x = y$, $Uc = d$. Then (24) is equivalent to $[S \ \ 0]y = d$. If $y = [y_1, y_2, \ldots, y_n]^T$, $d = [d_1, d_2, \ldots, d_r]^T$, then the variables $y_{r+1}, y_{r+2}, \ldots, y_n$ may be chosen in arbitrary fashion, but the variables $y_1, y_2, \ldots, y_r$ must satisfy

(25) $$s_i y_i = d_i, \qquad 1 \leq i \leq r$$

Provided that (25) has a solution, all the solutions of (24) may be found in terms of the $n - r$ parameters $y_{r+1}, y_{r+2}, \ldots, y_n$.

The same remarks apply if (24) is replaced by

(26) $$Ax \equiv c \bmod m$$

where m is any nonzero integer.

As an example, choose

$$A = \begin{bmatrix} 1 & 2 & 3 \\ 5 & 0 & 1 \end{bmatrix}, \qquad C = [-1, 11]^{\mathrm{T}}$$

It is readily determined that if

$$U = \begin{bmatrix} 1 & 0 \\ -5 & 1 \end{bmatrix}, \qquad V = \begin{bmatrix} 1 & 1 & -1 \\ 0 & 4 & -7 \\ 0 & -3 & 5 \end{bmatrix}$$

then

$$UAV = \begin{bmatrix} 1 & 0 & 0 \\ 0 & 2 & 0 \end{bmatrix} \text{ is in Smith normal form.}$$

Then $d = Uc = [-1, 16]^{\mathrm{T}}$. Thus (25) becomes

$$1 \cdot y_1 = -1, \qquad 2 \cdot y_2 = 16$$

so that

$$y_1 = -1, \qquad y_2 = 8$$

Therefore

$$x = Vy = V[-1, 8, y_3]^{\mathrm{T}} = [7 - y_3, 32 - 7y_3, -24 + 5y_3]^{\mathrm{T}}$$

Thus the complete solution in integers of the system

$$\begin{aligned} x_1 + 2x_2 + 3x_3 &= -1, \\ 5x_1 \qquad + x_3 &= \ \ 11 \end{aligned}$$

is given by

$$\begin{aligned} x_1 &= 7 - y_3, \\ x_2 &= 32 - 7y_3, \\ x_3 &= -24 + 5y_3 \end{aligned}$$

where y_3 is any integer.

(d) Let A be an $m \times n$ matrix with elements from a principal ideal ring R. Assume that $m < n$. We shall show that A may be completed to an $n \times n$ matrix B such that $\det(B)$ is an associate of $d_m(A)$.

Choose U in R_m', V in R_n' so that $UAV = C$ is in Smith normal form. Then $C = [S \ 0]$, $S = \operatorname{diag}(s_1, s_2, \ldots, s_m)$, and $s_1 s_2 \cdots s_m$ is an associate of $d_m(A)$. Complete C to the $n \times n$ matrix $S \dotplus I_{n-m}$. Then

$\det(C) = \det(S) = s_1 s_2 \cdots s_m$. Now define

$$B = (U^{-1} \dotplus I_{n-m})(S \dotplus I_{n-m})V^{-1}$$

Then $\det(B)$ is an associate of $d_m(A)$ and

$$B = \begin{bmatrix} A \\ E \end{bmatrix}, \qquad E = [0 \ \ I_{n-m}]V^{-1}$$

(e) Given objects $O_1, O_2, \ldots, O_p$ and sets $S_1, S_2, \ldots, S_q$ containing these objects, put $a_{ij} = 1$ if O_i belongs to S_j, and 0 otherwise. The $p \times q$ matrix $A = (a_{ij})$ so obtained is known as an *incidence* matrix. A fundamental combinatorial problem is to determine when a given incidence matrix can be obtained from another by permuting rows and columns. Thus if A, B are $p \times q$ incidence matrices, we say that A and B are *permutation equivalent* if a $p \times p$ permutation matrix P and a $q \times q$ permutation matrix Q exist such that $B = PAQ$. The problem of determining whether or not A and B are permutation equivalent is clearly a finite one, since it is only necessary to form all $p!\,q!$ products PAQ and check to see if B is among them; but this straightforward procedure is not in general feasible, because of the size of $p!\,q!$. Clearly, if A and B are not equivalent then they cannot be permutation equivalent. Thus a useful negative criterion is that if A and B do not have the same invariant factors (apart from unit multipliers) then they are not permutation equivalent.

EXERCISES AND PROBLEMS

In the following, R is a principal ideal ring.

1. Let $A \in R_m$, $B \in R_n$. Show that the elementary divisors of $A \dotplus B$ are the elementary divisors of A together with the elementary divisors of B.

2. Let $A \in R_m$, $B \in R_n$. Let π be any prime, and let $\pi^{k_1}, \pi^{k_2}, \ldots, \pi^{k_r}$ be the elementary divisors of A which are powers of π; and $\pi^{l_1}, \pi^{l_2}, \ldots, \pi^{l_s}$ the elementary divisors of B which are powers of π. Then the elementary divisors of $A \otimes B$ (the Kronecker product of A and B) which are powers of π are

$$\pi^{k_i + l_j}, \qquad 1 \leq i \leq r, \qquad 1 \leq j \leq s$$

Hint: First prove that $A \otimes B \,\tilde{E}\, S(A) \otimes S(B)$ and then use Theorem II.13.

3. Let $A, B \in R_n$ and suppose that $AB = \mu I$, $\mu \neq 0$. Then the invariant factors of A and of B are divisors of μ.

4. Let $A \in R_n$, A nonsingular. If $S(A) = \operatorname{diag}(s_1, s_2, \ldots, s_n)$, then apart from unit multipliers,

$$S(A^{\text{adj}}) = \operatorname{diag}(s/s_n, s/s_{n-1}, \ldots, s/s_1)$$

where $s = s_1 s_2 \cdots s_n$ is an associate of $\det(A)$.

5. Let $A \in Z_n$, A nonsingular. Then the least positive integer t such that $tA^{-1} \in Z_n$ is $t = |s_n(A)|$.

6. Let $A \in Z_n$, and suppose that $A^{k-1} + A^{k-2} + \cdots + I = 0$. Then the invariant factors of $A - I$ are divisors of k.

7. Let $C \in Z_n$,

$$C = \begin{bmatrix} 0 & 1 & 0 & \cdots & 0 \\ 0 & 0 & 1 & \cdots & 0 \\ \cdot & \cdot & \cdot & & \cdot \\ \cdot & \cdot & \cdot & & \cdot \\ \cdot & \cdot & \cdot & & \cdot \\ 0 & 0 & 0 & \cdots & 1 \\ a_1 & a_2 & a_3 & \cdots & a_n \end{bmatrix}$$

Then

$$S(I - C) = I_{n-1} \dotplus (a), \qquad a = a_1 + a_2 + \cdots + a_n$$

8. Let $A \in R_n$ have square-free determinant Δ (that is, not divisible by the square of any prime). Then apart from unit multipliers,

$$S(A) = I_{n-1} \dotplus (\Delta)$$

9. Let $A \in Z_3$, where

$$A = \begin{bmatrix} 2 & 6 & -8 \\ 12 & 14 & 6 \\ 4 & -4 & 8 \end{bmatrix}$$

Show that

$$A \ \tilde{E} \ \operatorname{diag}(2, 2, 168)$$

Chapter III

Similarity

A much more restrictive requirement than equivalence, similarity forms the basis for much of classical matrix theory.

1. Definition of similarity

Let A, B be elements of R_n. We say that A is *similar* to B (written $A \tilde{S} B$) if for some unit matrix U of R_n, $A = U^{-1}BU$. Similarity is thus an equivalence relationship which is a special type of two-sided equivalence, but has the property of being structure preserving, since for a given U of R_n' the mapping $A \longrightarrow U^{-1}AU$, $A \in R_n$, defines an inner automorphism of R_n.

2. A general result for any principal ideal ring

The general question of determining unique canonical forms with respect to similarity is unsolved. When R is a field, however, we can give a complete answer to this question. But first we prove a result for any principal ideal ring R.

Suppose that $1 \leq p < q \leq n$, and that x, y, u, v are elements of R such that $xv - yu = 1$. We set

$$M_{pq} = M_{pq}(x, y, u, v) \tag{1}$$
$$= I + (x-1)E_{pp} + yE_{pq} + uE_{qp} + (v-1)E_{qq}$$

Thus M_{pq} is the matrix which has x in the (p, p) position, v in the (q, q) position, 1 elsewhere on the diagonal, y in the (p, q) position, u in the (q, p) position, and 0 elsewhere off the diagonal.
M_{pq} has determinant 1, and

$$M_{pq}^{-1} = M_{pq}(v, -y, -u, x) \tag{2}$$

The theorem we wish to prove is the following:

Theorem III.1. Let A be any matrix of R_n. Then by an appropriate sequence of similarities using the matrices M_{pq}, A can be reduced to the form

$$C = \begin{bmatrix} c_{11} & c_{12} & 0 & \cdots & 0 \\ c_{21} & c_{22} & c_{23} & \cdots & 0 \\ \cdot & \cdot & \cdot & & \cdot \\ \cdot & \cdot & \cdot & & \cdot \\ \cdot & \cdot & \cdot & & \cdot \\ c_{n1} & c_{n2} & c_{n3} & \cdots & c_{nn} \end{bmatrix}$$

That is $A \tilde{S} C$, where $c_{ij} = 0$ if $j \geq i + 2$.

Proof. Let $2 < q \leq n$. We first note that the first row of $M_{2q}^{-1}A$ is the same as the first row of A. An easy computation shows that the first row of $M_{2q}^{-1}AM_{2q}$ is

$$[a_{11}, a_{12}x + a_{1q}u, a_{13}, \ldots, a_{1,q-1}, a_{12}y + a_{1q}v, a_{1,q+1}, \ldots, a_{1n}]$$

Suppose that in the matrix A, $a_{13} = a_{14} = \cdots = a_{1,q-1} = 0$, but $a_{1q} \neq 0$. Put $(a_{12}, a_{1q}) = \delta \neq 0$ and set $y = a_{1q}/\delta$, $v = -a_{12}/\delta$. Then $(y, v) = 1$ and hence x, u may be determined so that $xv - yu = 1$. With these values of x, y, u, v all the elements of $M_{2q}^{-1}AM_{2q}$ in positions $(1, 3)$, $(1,4), \ldots, (1, q)$ are zero.

It follows that if the transformations are chosen in the indicated manner and used in the order

$$M_{23}, M_{24}, \ldots, M_{2n}, M_{34}, \ldots, M_{3n}, \ldots, M_{n-1, n}$$

then the application of M_{pq} will bring element $(p-1, q)$ to zero and will not disturb the zeros previously produced. This completes the proof.

The matrix C is certainly not unique.

3. Similarity over a field

The theory of similarity for matrices with elements from a field F is best handled by means of the "λ-matrices," that is, matrices over $F[\lambda]$. Accordingly, we choose $R = F[\lambda]$ in what follows. Then the unit matrices of R_n are those whose determinants are the nonzero elements of F.

4. Degree and proper degree

Let A be any nonzero element of R_n. Since R_n may also be regarded as $F_n[\lambda]$, that is, as the ring of polynomials in λ over F_n, A has an expression

$$A = A_k\lambda^k + A_{k-1}\lambda^{k-1} + \cdots + A_0$$

where $A_0, A_1, \ldots, A_k$ belong to F_n and $A_k \neq 0$. We say that A_k is the *leading coefficient* of A, and that A is of *degree* k. If in addition A_k is nonsingular, we say that A is of *proper degree* k. We do not assign a degree to the zero matrix.

Some care is necessary in handling this notion, since AB may be of lower degree than either A or B. For example, $A = \left[\begin{smallmatrix} 0 & 1 \\ 0 & 0 \end{smallmatrix}\right]\lambda$ is of degree 1 but A^2 is the zero matrix. However, if A (say) is of proper degree k, then the degree of AB is the sum of the degrees of A and B.

5. A modified euclidean algorithm

The theorems that follow provide a modified euclidean algorithm for R_n.

Theorem III.2. Let A, B belong to R_n, and let B be of proper degree t. Then there are unique matrices Q, T of R_n such that $A = BQ + T$, and either $T = 0$ or $\deg T < t$.

Proof. If $A = 0$, then $Q = T = 0$ will do. Suppose then that $A \neq 0$.

Let

$$A = A_0\lambda^{k_0} + \cdots, \qquad A_0 \neq 0$$
$$B = B_0\lambda^t + \cdots, \qquad B_0 \text{ nonsingular}$$

If $k_0 < t$, set $Q = 0$, $T = A$. If $k_0 \geq t$, set $Q_1 = B_0^{-1}A_0\lambda^{k_0-t}$, $T_1 = A - BQ_1$. Then either $T_1 = 0$, or $\deg T_1 < \deg A$. If $T_1 = 0$ or $\deg T_1 < t$, set $Q = Q_1$, $T = T_1$. If $\deg T_1 = k_1 \geq t$, write $T_1 = A_1\lambda^{k_1} + \cdots$ and repeat the process, defining $Q_2 = B_0^{-1}A_1\lambda^{k_1-t}$, $T_2 = T_1 - BQ_2 = A - B(Q_1 + Q_2)$. Then either $T_2 = 0$, or $\deg T_2 < \deg T_1$. If $T_2 = 0$ or $\deg T_2 < t$, set $Q = Q_1 + Q_2$, $T = T_2$. If not, repeat the process, which must eventually terminate, since the degrees strictly decrease. Thus we have shown the existence of at least one pair Q, T.

To show the uniqueness, suppose that $A = BQ + T = BQ_0 + T_0$, where Q_0, T_0 also satisfy the hypotheses of the theorem. Then $B(Q - Q_0) = T_0 - T$. Suppose that $Q - Q_0 \neq 0$. Then, because B is of proper degree t, $\deg B(Q - Q_0) \geq t$. However, $T - T_0$ is either 0 or else of degree less than t. This is not possible, and hence $Q = Q_0$, $T = T_0$.

In just the same way, we have

Theorem III.3. Let A, B belong to R_n, and let B be of proper degree t. Then there are unique matrices Q_1, T_1 of R_n such that $A = Q_1B + T_1$, and either $T_1 = 0$, or $\deg T_1 < t$.

6. Two lemmas on degree and proper degree

We now prove two technical lemmas, which will lead to a theorem that settles the similarity question for a field completely.

Lemma III.1. Let $A, B \in R_n$, A of proper degree k, B of proper degree t. Suppose that there are matrices P_1, P_2 of R_n such that $AP_1 = P_2B$. Then there are matrices Q, T_1, T_2, of R_n such that $P_1 = QB + T_1$, $P_2 = AQ + T_2$, where $T_1 = 0$ or $\deg T_1 < t$, and $T_2 = 0$ or $\deg T_2 < k$. Furthermore, $AT_1 = T_2B$.

Proof. By Theorems III.2 and III.3, we can find $Q_1, Q_2, T_1, T_2 \in R_n$ such that $P_1 = Q_1B + T_1$, $P_2 = AQ_2 + T_2$, $T_1 = 0$ or $\deg T_1 < t$, $T_2 = 0$ or $\deg T_2 < k$. It follows that

$$AT_1 - T_2B = AP_1 - AQ_1B - P_2B + AQ_2B = A(Q_2 - Q_1)B$$

The left side is 0 or of degree less than $k + 1$, and the right side is 0 or of degree at least $k + t$. Therefore, both sides are 0, and so $AT_1 = T_2B$. Since A and B are of proper degrees k and t, respectively, $A(Q_2 - Q_1)B = 0$ implies that $Q_2 = Q_1$. This completes the proof, with $Q = Q_1 = Q_2$.

Lemma III.2. Let A, B, be equivalent matrices of R_n, which are of proper degree 1. Then there are non-singular matrices P, Q with elements in F such that $A = PBQ$.

Proof. Since $A \tilde{E} B$, there are unit matrices P_1, P_2 of R_n such that $A = P_2BP_1^{-1}$, $AP_1 = P_2B$. Determine matrices Q, T_1, T_2 of R_n as in Lemma IV.1. Then $AT_1 = T_2B$, and T_1, T_2 have elements in F. We need only prove that T_1, T_2 are nonsingular.

We may write $P_1^{-1} = Q_3A + T_3$, where $T_3 = 0$ or $\deg T_3 = 0$, so that T_3 has elements in F. Then

$$\begin{aligned} I = P_1^{-1}P_1 &= (Q_3A + T_3)(QB + T_1) \\ &= Q_3AQB + Q_3AT_1 + T_3QB + T_3T_1 \\ &= Q_3AQB + Q_3T_2B + T_3QB + T_3T_1, \\ I - T_3T_1 &= (Q_3AQ + Q_3T_2 + T_3Q)B \end{aligned}$$

Now the left side is 0 or of degree 0, and the right side is 0 or of degree at least 1. Hence both sides are 0 and so $I = T_3T_1$. Therefore, T_1 is nonsingular. Thus AT_1 has proper degree 1 and so T_2B has proper degree 1. Thus T_2 is nonsingular as well. This concludes the proof.

7. The fundamental theorem on similarity over a field

The next theorem completely settles the problem of similarity over a field F by reducing it to equivalence over $F[\lambda]$.

Theorem III.4. Let A, B be matrices of F_n. Then A and B are similar over F if and only if $\lambda I - A$ and $\lambda I - B$ are equivalent over $F[\lambda]$.

Proof. Suppose first that A and B are similar over F. Then there is a nonsingular matrix P of F_n such that $A = P^{-1}BP$. Hence $\lambda I - A = P^{-1}(\lambda I - B)P$, so that $\lambda I - A$ and $\lambda I - B$ are equivalent over $F[\lambda]$ (in fact, over F). Now suppose that $\lambda I - A$ and $\lambda I - B$ are equivalent over $F[\lambda]$. Since these are of proper degree 1, Lemma III.2 implies that nonsingular matrices P, Q of F_n exist such that $\lambda I - A = P(\lambda I - B)Q$. But then $PQ = I$, $PBQ = A$. Thus A and B are similar over F. This concludes the proof.

8. Minimal polynomials and nonderogatory matrices

It is well known that if A belongs to F_n, then $s_n(\lambda I - A)$ (the nth invariant factor of $\lambda I - A$) is the minimal polynomial $m(\lambda)$ of A, that is, the monic polynomial over F of least degree which annihilates A. The characteristic polynomial $d(\lambda)$ of A is of course given by

$$d(\lambda) = \det(\lambda I - A) = \prod_{k=1}^{n} s_k(\lambda I - A)$$

A matrix A for which the characteristic polynomial and minimal polynomial coincide is said to be *nonderogatory*. Thus A is nonderogatory if and only if

$$\lambda I - A \mathrel{\tilde{E}} I_{n-1} \dotplus (m(\lambda)) \tag{3}$$

(3) now implies the standard result:

Theorem III.5. A matrix A of F_n is nonderogatory if and only if the elementary divisors of $\lambda I - A$ consist of distinct prime powers.

9. Companion matrices

Let $g(\lambda) = \lambda^n + a_1\lambda^{n-1} + \cdots + a_n$ be any monic polynomial over F. The matrix B of F_n given by

$$B = \begin{bmatrix} 0 & 1 & 0 & \cdots & 0 \\ 0 & 0 & 1 & \cdots & 0 \\ \cdot & \cdot & \cdot & & \cdot \\ \cdot & \cdot & \cdot & & \cdot \\ \cdot & \cdot & \cdot & & \cdot \\ 0 & 0 & 0 & \cdots & 1 \\ -a_n & -a_{n-1} & -a_{n-2} & \cdots & -a_1 \end{bmatrix}$$

is known as the *companion matrix* of $g(\lambda)$. We have

Theorem III.6. Over $F[\lambda]$,

$$\lambda I - B \mathrel{\tilde{E}} I_{n-1} \dotplus (g(\lambda))$$

Thus B is nonderogatory.

Proof. For each k, $1 \leq k \leq n-1$, consider the $k \times k$ submatrix of $\lambda I - B$ obtained by choosing rows $1, 2, \ldots, k$ and columns $2, 3, \ldots, k+1$. Its diagonal elements are all -1, the elements on the subdiagonal are all λ, and the remaining elements are 0. The determinant of this matrix is thus $(-1)^k$. Therefore $d_k(\lambda I - B) = 1$, $1 \leq k \leq n-1$; and so the invariant factors $s_k(\lambda I - B)$ are all 1, $1 \leq k \leq n-1$. In addition, it is easily seen that $\det(\lambda I - B) = g(\lambda)$. This completes the proof.

This now implies

Theorem III.7. Every nonderogatory matrix over F is similar to the companion matrix of its minimum polynomial.

Proof. Let $A \in F_n$, and suppose that A is nonderogatory with minimal polynomial $m(\lambda)$. Let B be the companion matrix of $m(\lambda)$. Then $\lambda I - A \,\tilde{E}\, I_{n-1} \dotplus (m(\lambda))$; and by Theorem III. 6, $\lambda I - B \,\tilde{E}\, I_{n-1} + m(\lambda))$. Thus $\lambda I - A \,\tilde{E}\, \lambda I - B$, and hence $A \,\tilde{S}\, B$, by Theorem III.4. This completes the proof.

10. The Frobenius normal form

The next theorem is one of the classical results on canonical forms for similarity over a field.

Theorem III.8 (Frobenius normal form). Let A be any element of F_n and let $e_1(\lambda), e_2(\lambda), \ldots, e_k(\lambda)$ be the elementary divisors of $\lambda I - A$. Let B_i be the companion matrix of $e_i(\lambda)$, $1 \leq i \leq k$. Put

$$B = B_1 \dotplus B_2 \dotplus \cdots \dotplus B_k$$

Then $A \,\tilde{S}\, B$.

Proof. We have by Theorem III.6 that

$$\lambda I - B_i \,\tilde{E}\, \operatorname{diag}(1, 1, \ldots, 1, e_i(\lambda)), \qquad 1 \leq i \leq k$$

Thus $\lambda I - B$ is an $n \times n$ matrix with elementary divisors $e_1(\lambda)$, $e_2(\lambda), \ldots, e_k(\lambda)$. It follows that $\lambda I - A \,\tilde{E}\, \lambda I - B$, and so $A \,\tilde{S}\, B$, by Theorem III.4. This concludes the proof.

This normal form is also referred to as the *rational canonical form*.

11. Jordan matrices

When the field F is algebraically closed, other normal forms are possible which involve the Jordan matrices. If a is any element of F, the $n \times n$ matrix

$$J = J(a) = \begin{bmatrix} a & 1 & 0 & \cdots & 0 \\ 0 & a & 1 & \cdots & 0 \\ \cdot & \cdot & \cdot & & \cdot \\ \cdot & \cdot & \cdot & & \cdot \\ \cdot & \cdot & \cdot & & \cdot \\ 0 & 0 & 0 & \cdots & 1 \\ 0 & 0 & 0 & \cdots & a \end{bmatrix}$$

is known as a *Jordan matrix*. These matrices are nonderogatory. For the characteristic polynomial of J is $(\lambda - a)^n$ and so the minimal polynomial of J must be of the form $(\lambda - a)^k$, for some k such that $1 \leq k \leq n$. But the least positive integer k such that $(J - aI)^k = 0$ is n. Thus J is nonderogatory, and consequently

$$\lambda I - J \tilde{E} I_{n-1} \dotplus ((\lambda - a)^n)$$

12. The Jordan normal form

If F is algebraically closed and A belongs to F_n, then the elementary divisors of $\lambda I - A$ must be of the form $(\lambda - a)^k$, since the primes of $F[\lambda]$ are the linear polynomials, which may be taken to be monic.
We have

Theorem III.9 (Jordan normal form). Let A belong to F_n where F is algebraically closed. Let $p_i(\lambda) = (\lambda - \lambda_i)^{e_i}$, $1 \leq i \leq k$, denote the elementary divisors of $\lambda I - A$, where the elements $\lambda_1, \lambda_2, \ldots, \lambda_k$ belong to F, but are not necessarily distinct. Let J_i be the Jordan matrix of order e_i corresponding to the elementary divisor $p_i(\lambda)$, $1 \leq i \leq k$, and let

$$B = J_1 \dotplus J_2 \dotplus \cdots \dotplus J_k$$

Then $A \tilde{S} B$.

Proof. The matrix $J_i - \lambda I$ is also a Jordan matrix and therefore nonderogatory, which implies that $\lambda I - J_i$ has $p_i(\lambda)$ as its only elementary divisor, $1 \leq i \leq k$. Thus $\lambda I - A$ and $\lambda I - B$ are each $n \times n$ matrices with the same elementary divisors. It follows that $\lambda I - A \,\tilde{E}\, \lambda I - B$, and so $A \,\tilde{S}\, B$, by Theorem III.4. This completes the proof.

13. Criteria for similarity to a diagonal matrix

The next two theorems are famous classical criteria for similarity to a diagonal matrix

Theorem III.10 (Weierstrass). Let A belong to F_n, where F is algebraically closed. Then A is similar to a diagonal matrix over F if and only if $\lambda I - A$ has simple elementary divisors.

Proof. Suppose that $\lambda I - A$ has simple elementary divisors. Define J_i, $1 \leq i \leq k$, as in the preceding theorem. Then each J_i must in fact be $[\lambda_i]$, $1 \leq i \leq k$. Hence $k = n$, and $A \,\tilde{S}\, \text{diag}(\lambda_1, \lambda_2, \ldots, \lambda_n)$. Conversely, suppose that A is similar to the diagonal matrix $\text{diag}(\lambda_1, \lambda_2, \ldots, \lambda_n)$. Then $\lambda I - A$ is similar to $\text{diag}(\lambda - \lambda_1, \lambda - \lambda_2, \ldots, \lambda - \lambda_n)$, so that $\lambda I - A \,\tilde{E}\, \text{diag}(\lambda - \lambda_1, \lambda - \lambda_2, \ldots, \lambda - \lambda_n)$. Hence the elementary divisors of $\lambda I - A$ are simple. This completes the proof.

An immediate consequence of the above, whose proof we omit, is

Theorem III.11. If A belongs to F_n, where F is algebraically closed, then A is similar to a diagonal matrix over F if and only if the zeros of its minimal polynomial are distinct.

14. Similarity over Z

The theorems just developed may be regarded as a complete answer to the question of similarity over a field. As was mentioned earlier, the general problem is still unsolved for an arbitrary principal ideal ring R. The most important special case when R is not a field is undoubtedly the case $R = Z$. Although no complete solution can be given, the problem has been reduced to the determination of the ideal classes in certain rings of algebraic integers.

We must presuppose an acquaintance with the elements of algebraic number theory in what follows. A good general reference is Hecke's book [16], or the monograph [52].

The rational field will be denoted by Q. If K is an algebraic number field of degree n over Q, a module basis for the integers of K will be denoted by $\omega_1, \omega_2, \ldots, \omega_n$, and a generator of K by θ, so that $K = Q(\theta)$. We may always choose θ to be an algebraic integer of K.

The ring of integers of K, which coincides with the totality $\{x_1\omega_1 + x_2\omega_2 + \cdots + x_n\omega_n\}$ where $x_1, x_2, \ldots, x_n$ run over Z independently, will be denoted by K^*.

We go on to the problem of determining the similarity classes of matrices of Z_n. Since the characteristic polynomial is a similarity invariant, a first requirement is that matrices in the same similarity class must have the same characteristic polynomial. It is sufficient therefore to restrict ourselves to the matrices of Z_n having some given characteristic polynomial $f(\lambda)$.

Clearly, similarity over Z implies similarity over Q. It follows that a necessary condition for the matrices A, B of Z_n to be similar over Z is that $\lambda I - A$ and $\lambda I - B$ have the same invariant factors over $Q[\lambda]$.

When $f(\lambda)$ is irreducible over Q, the situation is somewhat simpler. For if A, B are matrices of Z_n with characteristic polynomial $f(\lambda)$, then the minimal polynomial of each of them must also be $f(\lambda)$, so that A and B are nonderogatory. Thus $\lambda I - A$ and $\lambda I - B$ are each equivalent over $Q[\lambda]$ to $I_{n-1} \dot{+} (f(\lambda))$. It follows that A and B are similar over Q.

15. Similarity to a block triangular matrix

The theorem that follows gives a partial reduction of the similarity problem to the case of an irreducible characteristic polynomial:

Theorem III.12. Let A be any element of Z_n. Then A is similar over Z to a block triangular matrix

$$\begin{bmatrix} A_{11} & A_{12} & \cdots & A_{1p} \\ 0 & A_{22} & \cdots & A_{2p} \\ \cdot & \cdot & & \cdot \\ \cdot & \cdot & & \cdot \\ \cdot & \cdot & & \cdot \\ 0 & 0 & \cdots & A_{pp} \end{bmatrix}$$

where the characteristic polynomial of A_{ii} is irreducible over Q, $1 \leq i \leq p$.

Proof. Let $f(\lambda)$ be the characteristic polynomial of A, so that $f(\lambda)$ is monic and has integral coefficients. Let θ be an eigenvalue of A. Then for some monic irreducible factor $g(\lambda)$ of $f(\lambda)$, $g(\theta) = 0$. Suppose that $\deg g(\lambda) = k$, and let $\omega_1, \omega_2, \ldots, \omega_k$ be a basis for K^*, $K = Q(\theta)$. Then if x is any eigenvector of A corresponding to the eigenvalue θ, the coordinates of x may be chosen as elements of K^*. Choose such a nonzero x. Then for some integral $n \times k$ matrix C,

$$x = C\omega, \qquad \omega = [\omega_1, \omega_2, \ldots, \omega_k]^{\mathrm{T}}$$

Also for some integral $k \times k$ matrix B,

$$\theta\omega = B\omega$$

Thus from $Ax = \theta x$ we get that $AC\omega = CB\omega$, so that

$$AC = CB$$

since $AC - CB$ is an integral matrix and the elements of ω are linearly independent over the rationals.

Suppose that C is of rank r. Then $r \leq k$, since C is an $n \times k$ matrix. Let

$$C = U\begin{bmatrix} S & 0 \\ 0 & 0 \end{bmatrix}V$$

where $U \in Z_n'$, $V \in Z_k'$, and S is a nonsingular $r \times r$ matrix which may be taken in Smith normal form, although this is not essential. Then if

$$U^{-1}AU = \begin{bmatrix} A_1 & A_2 \\ A_3 & A_4 \end{bmatrix}$$

where A_1 is $r \times r$,

$$\begin{bmatrix} A_1 & A_2 \\ A_3 & A_4 \end{bmatrix}\begin{bmatrix} S & 0 \\ 0 & 0 \end{bmatrix}V = \begin{bmatrix} S & 0 \\ 0 & 0 \end{bmatrix}VB$$

which implies that $A_3 = 0$. Now multiply by ω on the right to get

$$\begin{bmatrix} A_1 & A_2 \\ 0 & A_4 \end{bmatrix}\begin{bmatrix} S & 0 \\ 0 & 0 \end{bmatrix}V\omega = \begin{bmatrix} S & 0 \\ 0 & 0 \end{bmatrix}VB\omega = \theta\begin{bmatrix} S & 0 \\ 0 & 0 \end{bmatrix}V\omega$$

Put $\tau = V\omega$. Then the entries of τ are also linearly independent over the rationals. Write $\tau = \left[\begin{smallmatrix} \alpha \\ \beta \end{smallmatrix}\right]$, where α is $r \times 1$. Then

$$A_1 S\alpha = \theta S\alpha$$

This implies that θ is an eigenvalue of A_1, since $S\alpha \neq 0$. Hence $r = k$ and the characteristic polynomial of A_1 is $g(\lambda)$, since $g(\lambda)$ is irreducible.

We have proved therefore that A is similar over Z to

$$\begin{bmatrix} A_1 & A_2 \\ 0 & A_4 \end{bmatrix}$$

where the characteristic polynomial of A_1 is $g(\lambda)$. The result now follows by repeating the argument until all the irreducible factors of the characteristic polynomial have been used. This concludes the proof.

16. The theorem of Latimer and MacDuffee

We now go on to the case when $f(\lambda)$ is a monic polynomial of degree n over Z which is irreducible over the rationals. The discussion that follows is due to Taussky [67], and is her proof of the general theorem of Latimer and MacDuffee for the case of an irreducible polynomial.

Let θ be a root of $f(\lambda)$, and set $R = Z[\theta]$. Then $1, \theta, \ldots, \theta^{n-1}$ is a module basis for R. Let A be any $n \times n$ matrix over Z such that $f(A) = 0$. Since $f(\lambda)$ is irreducible and of degree n, it must be both the characteristic and minimum polynomial of A. It follows that θ is an eigenvalue of A and that the eigenvalues of A are distinct. Thus any two eigenvectors of A corresponding to the eigenvalue θ must be proportional.

Since A has at least one eigenvector corresponding to θ whose coordinates lie in $Q(\theta)$, it must have one, say x, whose coordinates lie in R; and any other eigenvector of A corresponding to θ must be a multiple of x.

Let $x = [x_1, x_2, \ldots, x_n]^{\mathrm{T}}$, and let S be the Z-module generated by $x_1, x_2, \ldots, x_n$. Since $\theta x = Ax$ and the coordinates of Ax belong to S, $\theta S \subset S$. It follows that $\theta^k S \subset S$ for any nonnegative integer k and hence that $\alpha S \subset S$ for every element α of R, since $1, \theta, \ldots, \theta^{n-1}$ is a module basis for R. Thus S is an ideal of R, determined by x and hence by A. In addition any other eigenvector y of A corresponding to θ determines an ideal in the same class as S, since x and y are proportional (two ideals S and T of R are said to be in the same ideal class, if nonzero elements α, β of R exist such that $\alpha S = \beta T$). Hence A uniquely determines an ideal class of R. Furthermore, if B is similar to

A over Z_n, so that $B = UAU^{-1}$ for some U belonging to Z_n', then B determines the same ideal class as A, since Ux is an eigenvector of B corresponding to θ, and the ideals determined by x and Ux are the same.

Now let $\mathfrak{S}$ be any ideal class of R, and suppose that $S \in \mathfrak{S}$. Let $x_1, x_2, \ldots, x_n$ be a module basis for S, and put $x = [x_1, x_2, \ldots, x_n]^T$. Then for some $A \in Z_n$, $\theta x = Ax$. Hence $f(A)x = f(\theta)x = 0$, and so $f(A) = 0$, since $x_1, x_2, \ldots, x_n$ are linearly independent over Q. Thus every ideal class $\mathfrak{S}$ determines an element A of Z_n such that $f(A) = 0$.

Suppose now that A, B are matrices of Z_n which have been obtained this way: say A by the ideal $S \in \mathfrak{S}$ with module basis $x_1, x_2, \ldots, x_n$ and B by the ideal $T \in \mathfrak{S}$ with module basis $y_1, y_2, \ldots, y_n$. Then nonzero elements α, β of R exist such that $\alpha S = \beta T$. Put $x = [x_1, x_2, \ldots, x_n]^T$, $y = [y_1, y_2, \ldots, y_n]^T$. Then for some unit matrix U of Z_n, $\beta y = U\alpha x$. Since $Ax = \theta x$, $By = \theta y$, we get

$$\beta By = \beta\theta y = \theta U\alpha x = U\alpha Ax,$$

and also

$$\beta By = BU\alpha x$$

It follows that

$$UAx = BUx,$$
$$(A - U^{-1}BU)x = 0$$

But this implies that $A = U^{-1}BU$, since the coordinates of x are linearly independent over Q. Thus matrices derived from the same ideal class are similar over Z. Summarizing, we have proved

Theorem III.13. There is a one-to-one correspondence between the similarity classes of matrices A of Z_n such that $f(A) = 0$, and the ideal classes of the ring $R = Z[\theta]$, where $f(\lambda)$ is a monic polynomial of degree n with integral coefficients which is irreducible over Q and θ is a root of $f(\lambda)$.

As a most important consequence, using the fact that the number of ideal classes is finite, we get

Theorem III.14. There are only finitely many similarity classes of matrices A of Z_n such that $f(A) = 0$, where $f(\lambda)$ is a monic polynomial of degree n with integral coefficients which is irreducible over Q. The number of classes is the same as the class number of the ring $Z[\theta]$, where θ is any root of $f(\lambda)$.

For example, this theorem implies that every matrix A of Z_2 such that $A^2 + I = 0$ is similar over Z to $T = \left[\begin{smallmatrix} 0 & 1 \\ -1 & 0 \end{smallmatrix}\right]$, since the class number of $Z[i]$ is 1, and the matrix T satisfies $T^2 + I = 0$.

EXERCISES AND PROBLEMS

1. Suppose that A belongs to Z_n and that A has an eigenvalue a which belongs to Z. Prove that A is similar over Z to a matrix whose first column is $[a, 0, \ldots, 0]^{\mathrm{T}}$.

Outline of proof: Choose an eigenvector x of A belonging to the eigenvalue α so that the coordinates of x are relatively prime rational integers. Complete x to an $n \times n$ unit matrix U of Z_n; that is, the first column of U is x. Then show that the first column of $U^{-1}AU$ is $[a, 0, \ldots, 0]^{\mathrm{T}}$.

2. Let A belong to Z_n, $A^2 = I$. Show that A is similar over Z to one of the matrices

$$\begin{bmatrix} I_r & B_{r,n-r} \\ 0 & -I_{n-r} \end{bmatrix}, \qquad 0 \leq r \leq n$$

where the matrix $B_{r,n-r}$ is in Smith normal form and its elements may be taken modulo 2; and no two of these are similar over Z. Hence deduce that the total number of similarity classes is

$$\sum_{r=0}^{n} \{1 + \min(r, n-r)\}$$

3. Let $A = \left[\begin{smallmatrix} A_1 & A_2 \\ A_3 & A_4 \end{smallmatrix}\right]$ belong to R_n, where A_1 is square. Show that A_2 may always be taken in Smith normal form after an appropriate similarity by a unit matrix of R_n.

4. Show that every matrix A of Z_{2n} such that $A^2 + I = 0$ is similar to the direct sum of n copies of $\left[\begin{smallmatrix} 0 & 1 \\ -1 & 0 \end{smallmatrix}\right]$.

5. Show that every matrix A of Z_2 such that $A^2 + A + I = 0$ is similar over Z to $U = \left[\begin{smallmatrix} 0 & 1 \\ -1 & -1 \end{smallmatrix}\right]$.

6. Suppose that $A = \left[\begin{smallmatrix} A_1 & A_2 \\ 0 & A_4 \end{smallmatrix}\right]$ belongs to Z_t, $t = \varphi(m) + \varphi(n)$, where the characteristic polynomial of A_1 is the mth cyclotomic polynomial, and the characteristic polynomial of A_4 the nth cyclotomic polynomial. Assume that $m > 1$, $n > 1$, $(m, n) = 1$. Then A is similar over Z to $A_1 \dotplus A_4$.

Outline of proof. Perform the similarity

$$\begin{bmatrix} I & X \\ 0 & I \end{bmatrix} \begin{bmatrix} A_1 & A_2 \\ 0 & A_4 \end{bmatrix} \begin{bmatrix} I & -X \\ 0 & I \end{bmatrix} = \begin{bmatrix} A_1 & A_2 + XA_4 - A_1X \\ 0 & A_4 \end{bmatrix}$$

The system of linear equations

$$XA_4 - A_1X + A_2 = 0$$

is permutation equivalent to a system in which the coefficient matrix is

$$C = I \otimes A_1 - A_4 \otimes I$$

The eigenvalues of C are the differences of the eigenvalues of A_1 and A_2, and hence are all of the form $\zeta_m - \zeta_n$, where ζ_m is a primitive mth root of unity and ζ_n a primitive nth root of unity. But these are all units in the field of the mnth roots of unity, since mn is not a prime power. Hence C is unimodular, and so an integral matrix X may be determined which satisfies the original system, and brings A to the desired form.

7. Let $A = \begin{bmatrix} A_1 & A_2 \\ 0 & A_4 \end{bmatrix}$ belong to F_n, where A_1 is square. If A_1, A_4 have no eigenvalue in common (the eigenvalues of course lie in some finite extension of F), then A is similar over F to $A_1 \dotplus A_4$.

Outline of proof: As in Problem 6, the matrix $C = I \otimes A_1 - A_4 \otimes I$ is nonsingular, and hence X may be determined with coefficients from F so that

$$XA_4 - A_1X + A_2 = 0$$

which implies the result.

8. Give an example of a pair of matrices in Z_2 which are similar over Q but not over Z (one example: $\begin{bmatrix} 3 & 2 \\ 4 & 3 \end{bmatrix}$, $\begin{bmatrix} 0 & 1 \\ -1 & 6 \end{bmatrix}$).

9. Show that every matrix A of Z_2 such that $A^2 - tA + \Delta I = 0$, where $p(\lambda) = \lambda^2 - t\lambda + \Delta$ is irreducible over Q, is similar over Z to a matrix $B = \begin{bmatrix} a & b \\ c & d \end{bmatrix}$ such that $2|a| \leq |c| \leq |d|$.

Hint: $|c|$ may be characterized as the least value assumed by $|\gamma|$, where $A = \begin{bmatrix} \alpha & \beta \\ \gamma & \delta \end{bmatrix}$ runs over the set of matrices under consideration.

Chapter IV

Congruence

The material of this chapter finds its application in the theory of quadratic forms and is the basis of much of classical matrix theory and classical number theory.

1. Definition of congruence

Let A, B be elements of R_n. We say that A is *congruent* to B (written $A \tilde{C} B$) if there is a unit matrix U of R_n such that $A = U^{\mathrm{T}}BU$. Once again this is an equivalence relationship. It is not as strong as similarity since it does not define a homomorphism of R_n into R_n, but it is certainly more restrictive than equivalence.

2. Definition of quadratic form

Let $x = [x_1, x_2, \ldots, x_n]^{\mathrm{T}}$ be an $n \times 1$ vector with entries from R, and let $A = (a_{ij})$ be an element of R_n. Then the *quadratic form* associated with A is

$$q = x^{\mathrm{T}}Ax = \sum_{1 \leq i, j \leq n} a_{ij}x_ix_j$$

The transformation $A \longrightarrow U^{\mathrm{T}}AU$ corresponds to the substitution $x = Uy$, which replaces $x^{\mathrm{T}}Ax$ by the "congruent form" $y^{\mathrm{T}}U^{\mathrm{T}}AUy$. Since the value of q is unchanged if A is replaced by A^{T}, the study of congruence transformations is usually restricted to *symmetric* matrices (those for which $A^{\mathrm{T}} = A$) or *skew-symmetric* matrices (those for which $A^{\mathrm{T}} = -A$), and these properties are preserved under congruence. As with similarity, the general problem of deciding when two given matrices of R_n are congruent is unsolved. In fact, it has not even been solved completely when R is a field. However for the skew-symmetric matrices there is a complete and simple solution, provided that the characteristic of R is not 2.

We note that if an elementary row operation is performed on the rows of a matrix, and if the corresponding elementary column operation is then performed on the columns, the net result is the same as applying a congruence transformation, since the elementary column matrix corresponding to the elementary row matrix E is E^{T}.

3. The skew normal form

We first treat the skew-symmetric matrices.

Theorem IV.1 (skew normal form). Let A be a skew-symmetric matrix of rank r which belongs to R_n, where the principal ideal ring R is not of characteristic 2. Then $r = 2s$ and

$$A \tilde{C} \begin{bmatrix} 0 & h_1 \\ -h_1 & 0 \end{bmatrix} \dotplus \begin{bmatrix} 0 & h_2 \\ -h_2 & 0 \end{bmatrix} \dotplus \cdots \dotplus \begin{bmatrix} 0 & h_s \\ -h_s & 0 \end{bmatrix} \dotplus 0$$

where $h_i \mid h_{i+1}$, $1 \leq i \leq s - 1$.

Proof. We may assume that $A \neq 0$. Then a nonzero element may be brought to the first row by a permutation of the rows and a corresponding permutation of the columns. Therefore we may assume without loss of generality that the first row of A contains a nonzero element which cannot occur in the (1, 1) position, since the diagonal elements of any skew-symmetric matrix of R_n must all be 0. Define

$$(a_{12}, a_{13}, \ldots, a_{1n}) = \delta$$

and let $x_2, x_3, \ldots, x_n$ be elements of R such that

$$a_{12}x_2 + a_{13}x_3 + \cdots + a_{1n}x_n = \delta$$

Then certainly $\delta \neq 0$, and $(x_2, x_3, \ldots, x_n) = 1$. We may determine a unit matrix U of R_{n-1} whose first column is $[x_2, x_3, \ldots, x_n]^{\mathrm{T}}$. Then $V = I_1 \dotplus U$ belongs to R_n'. Put

$$B = V^{\mathrm{T}}AV = \begin{bmatrix} 0 & b_{12} & b_{13} & \cdots \\ -b_{12} & 0 & b_{23} & \cdots \\ -b_{13} & -b_{23} & 0 & \cdots \\ \cdot & \cdot & \cdot & \\ \cdot & \cdot & \cdot & \\ \cdot & \cdot & \cdot & \end{bmatrix}$$

Then a brief computation shows that $b_{12} = \delta$ and that $b_{12} \mid b_{1j}$, $3 \leq j \leq n$. Now add $-b_{1j}/b_{12}$ times row 2 to row j and then add $-b_{1j}/b_{12}$ time column 2 to column j, $3 \leq j \leq n$. The result is that A is congruent to a matrix whose first row is $[0, b_{12}, 0, \ldots, 0]$. The procedure is now repeated with the submatrix obtained by striking out the first row and column, until a triple diagonal matrix is reached, so that

$$A \tilde{C} C = \begin{bmatrix} 0 & c_{12} & 0 & 0 & \cdots \\ -c_{12} & 0 & c_{23} & 0 & \cdots \\ 0 & -c_{23} & 0 & c_{34} & \cdots \\ 0 & 0 & -c_{34} & 0 & \cdots \\ \cdot & \cdot & \cdot & & \\ \cdot & \cdot & \cdot & & \\ \cdot & \cdot & \cdot & & \end{bmatrix}$$

where $c_{12} = b_{12} \neq 0$. There are now two possibilities: either c_{12} divides every element of C or it does not. If not, add rows $2, 3, \ldots, n$ to row 1 and columns $2, 3, \ldots, n$ to column 1. The first row of the matrix so obtained is

$$[0, c_{12} - c_{23}, c_{23} - c_{34}, \ldots, c_{n-2,n-1} - c_{n-1,n}, c_{n-1,n}]$$

Now the greatest common divisor γ of the first row elements is readily seen to be

$$\gamma = (c_{12}, c_{23}, \ldots, c_{n-1,n})$$

and γ contains fewer prime divisors than c_{12}. As in the first part of the proof, c_{12} may now be replaced by γ and the entire process repeated until a triple diagonal matrix is obtained having γ in the (1, 2) position. Thus in a finite number of steps a triple diagonal matrix D congruent to

A may be obtained in which the (1, 2) element divides every element of the matrix. Let $D = (d_{ij})$, where now $d_{12} \mid d_{ij}$ for $1 \leq i, j \leq n$. Add d_{23}/d_{12} times row 1 to row 3, and then add d_{23}/d_{12} times column 1 to column 3. Then D is congruent to the direct sum

$$\begin{bmatrix} 0 & d_{12} \\ -d_{12} & 0 \end{bmatrix} \dotplus E$$

where d_{12} divides every element of E. The process is now repeated with E until the canonical form described by the theorem is obtained, and the procedure makes it clear that the rank of A must be even. This concludes the proof.

Theorem IV.2. The elements

$$h_1, h_1, h_2, h_2, \ldots, h_s, h_s$$

of the preceding theorem are the invariant factors of A, and hence the Skew normal form is unique.

Proof. It is only necessary to note that if h belongs to R, then

$$\begin{bmatrix} 0 & h \\ -h & 0 \end{bmatrix} \tilde{E} \begin{bmatrix} h & 0 \\ 0 & h \end{bmatrix},$$

since

$$\begin{bmatrix} 0 & h \\ -h & 0 \end{bmatrix} = \begin{bmatrix} 0 & 1 \\ -1 & 0 \end{bmatrix} \begin{bmatrix} h & 0 \\ 0 & h \end{bmatrix}$$

and hence

$$\begin{bmatrix} 0 & h_2 \\ -h_1 & 0 \end{bmatrix} \dotplus \begin{bmatrix} 0 & h_2 \\ -h_2 & 0 \end{bmatrix} \dotplus \cdots \dotplus \begin{bmatrix} 0 & h_s \\ -h_s & 0 \end{bmatrix} \dotplus 0$$
$$\tilde{E} \operatorname{diag}(h_1, h_1, h_2, h_2, \ldots, h_s, h_s) \dotplus 0$$

Furthermore $h_1 \mid h_1 \mid h_2 \mid h_2 \mid \cdots \mid h_s \mid h_s$.

We summarize some consequences of these theorems:

Theorem IV.3. Let A, B be skew-symmetric matrices of R_n, where R is not of characteristic 2. Then

(1) $A \tilde{C} B$ if and only if A and B have the same invariant factors.
(2) If $A \tilde{E} B$, then $A \tilde{C} B$.
(3) The $(2k-1)$th and $(2k)$th invariant factors of A are equal.

(4) The number of congruence classes of nonsingular, skew-symmetric matrices of R_{2n} of determinant $\epsilon\Delta^2$, ϵ a unit of R (the determinant must be an associate of a square) is finite and equal to $S_n(\Delta)$.

4. A general result for symmetric matrices

We now consider the symmetric matrices. Our first result is

Theorem IV.4. Let A belong to R_n, $A = A^T$. Then A is congruent to a triple diagonal matrix $S = (s_{ij})$, that is, one for which $s_{ij} = 0$ unless $i - j = 0, 1, -1$.

Proof. The proof is practically identical with the first part of the proof of Theorem IV.1, but much simpler. Consider the elements $a_{13}, a_{14}, \ldots, a_{1n}$. If they are not all 0, put $(a_{12}, a_{13}, \ldots, a_{1n}) = \delta$, so that $\delta \neq 0$. Then elements $x_2, x_3, \ldots, x_n$ of R exist such that $a_{12}x_2 + a_{13}x_3 + \cdots + a_{1n}x_n = \delta$, and $(x_2, x_3, \ldots, x_n) = 1$. Determine a unit matrix U of R_{n-1} whose first column is $[x_2, x_3, \ldots, x_n]^T$ and put $V = I_1 \dotplus U$, $B = V^T A V$. Then $b_{12} = \delta$, and δ divides b_{13}, $b_{14}, \ldots, b_{1n}$. But now by elementary row operations and corresponding elementary column operations, A may be brought to a form where all the elements of the first row and column with the possible exception of elements (1, 1), (1, 2), (2, 1), are 0. The process is now repeated with the submatrix obtained by deleting the first row and column until the desired form is reached.

This form is not unique. For the case when R is euclidean, the procedure can be carried out by elementary operations alone.

5. The connection between symmetric matrices and quadratic forms

At this point we outline the connection between symmetric matrices and quadratic forms. Although the classical theory deals with matrices over fields or over Z, the subject can be initiated for any principal ideal ring R.

6. The set $q(A)$

Let V_n denote the set of $n \times 1$ vectors over R. Let $A = (a_{ij})$ belong to R_n, $A = A^T$. Let $x = [x_1, x_2, \ldots, x_n]^T$ be any element of V_n. Then if

$$\alpha = x^{\mathrm{T}}Ax = \sum_{1 \leq i, j \leq n} a_{ij}x_ix_j$$

we say that A *represents* α. The set of values represented by A will be denoted by $q(A)$, so that

$$q(A) = \{x^{\mathrm{T}}Ax : x \in V_n\}$$

It is clear that $q(A)$ is a congruence invariant; for if $B = U^{\mathrm{T}}AU$, U a unit matrix of R_n, then

$$q(B) = q(U^{\mathrm{T}}AU) = \{x^{\mathrm{T}}U^{\mathrm{T}}AUx : x \in V_n\}$$

Put $y = Ux$. Then as x runs over V_n once and once only, so does y, since U is a unit matrix. Thus

$$q(B) = \{y^{\mathrm{T}}Ay : y \in V_n\} = q(A)$$

Unfortunately the converse is not necessarily true. Nevertheless the set $q(A)$ is a most important concept, and underlies the theory of congruence. Indeed, one of the classical problems of number theory is the determination of the set $q(A)$.

7. Results on elements represented by forms

The importance of the set becomes clear in view of the following lemma:

Lemma IV.1. Suppose that A represents $\alpha, \alpha = x^{\mathrm{T}}Ax, x \neq 0$. Put $(x_1, x_2, \ldots, x_n) = \delta$. Then α/δ^2 occurs as the (1, 1) element of some matrix B congruent to A.

Proof. Since $(x_1, x_2, \ldots, x_n) = \delta$, the vector $y = x/\delta$ belongs to V_n and has relatively prime coordinates. Let U be a unit matrix of R_n whose first column is y. Put $B = U^{\mathrm{T}}AU$. Let $e_1 = [1, 0, \ldots, 0]^{\mathrm{T}}$. Then $y = Ue_1$, $e_1{}^{\mathrm{T}}Be_1 = e_1{}^{\mathrm{T}}U^{\mathrm{T}}AUe_1 = y^{\mathrm{T}}Ay = \alpha/\delta^2$. But the (1, 1) element of B is $e_1{}^{\mathrm{T}}Be_1$. This completes the proof.

An important corollary is

Theorem IV.5. Suppose that α is represented by A and that α is either a nonzero element of $q(A)$ with a minimal number of prime divisors, or if R is euclidean, an element of $q(A)$ for which $v(\alpha)$ is minimal. Then α occurs as the (1, 1) element of some matrix B congruent to A.

Proof. Let $x \in V_n$ be such that

$$\alpha = x^{\mathrm{T}}Ax$$

Then certainly $x \neq 0$, since $\alpha \neq 0$. Put $(x_1, x_2, \ldots, x_n) = \delta$. Then α/δ^2 also belongs to $q(A)$. The minimality condition now implies that δ must be 1, since in the first case α/δ^2 would have fewer prime factors then α and in the second, $v(\alpha/\delta^2)$ would be less than $v(\alpha)$. But now the result follows from the preceding lemma.

8. Congruence over a field

We now choose $R = F$, a field. We first prove

Theorem IV.6. Let A be a symmetric matrix of rank r belonging to F_n. Then

$$A \ \tilde{C} \ \operatorname{diag}(d_1, d_2, \ldots, d_p) \dotplus qT \dotplus 0$$

where $d_i \neq 0$, $1 \leq i \leq p$, T is the matrix $\left[\begin{smallmatrix} 0 & 1 \\ 1 & 0 \end{smallmatrix}\right]$, qT denotes the direct sum of q copies of T, $r = p + 2q$, and q may be 0.

Proof. We may assume that the first row of A is not 0. There are two cases, leading to two different reduction processes:

(i) Suppose that some diagonal element of A is not 0. Then by a suitable row permutation followed by a corresponding column permutation, this diagonal element may be brought into the (1, 1) position. Thus we may assume that $a_{11} \neq 0$. Now subtract a_{1i}/a_{11} times row 1 from row i and then subtract a_{1i}/a_{11} times column 1 from column i, $2 \leq i \leq n$. This produces a matrix congruent to A which is the direct sum of (a_{11}) with an $(n-1) \times (n-1)$ matrix.

(ii) Suppose that every diagonal element of A is 0. Then $a_{1j} \neq 0$ for some j such that $2 \leq j \leq n$. Interchanging rows 2 and j and columns 2 and j brings this element into the (1,2) position and does not change the fact that all the diagonal elements are 0. We may assume then that A has the form

$$A = \begin{bmatrix} 0 & a_{12} & \cdots \\ a_{12} & 0 & \cdots \\ \cdot & \cdot & \\ \cdot & \cdot & \\ \cdot & \cdot & \end{bmatrix}$$

where $a_{12} \neq 0$. Put

$$U = \begin{bmatrix} 0 & a_{12}^{-1} \\ 1 & 0 \end{bmatrix} \dotplus I_{n-2}$$

Then U is a unit matrix of F_n and

$$B = UAU^{\mathrm{T}} = \begin{bmatrix} 0 & 1 & b_{13} & \cdots \\ 1 & 0 & b_{23} & \cdots \\ \cdot & \cdot & \cdot & \\ \cdot & \cdot & \cdot & \\ \cdot & \cdot & \cdot & \end{bmatrix}$$

Now subtract b_{1j} times row 2 from row j and then b_{1j} times column 2 from column j, $3 \leq j \leq n$, to produce the congruent matrix

$$C = \begin{bmatrix} 0 & 1 & 0 & \cdots & 0 \\ 1 & 0 & c_{23} & \cdots & c_{2n} \\ \cdot & \cdot & \cdot & & \cdot \\ \cdot & \cdot & \cdot & & \cdot \\ \cdot & \cdot & \cdot & & \cdot \end{bmatrix}$$

Finally, subtract c_{2j} times row 1 from row j and then c_{2j} times column 1 from column j, $3 \leq j \leq n$. This produces a matrix congruent to A which is the direct sum of $\left[\begin{smallmatrix} 0 & 1 \\ 1 & 0 \end{smallmatrix}\right]$ with an $(n-2) \times (n-2)$ matrix.

It is clear that repetition of these processes applied to the submatrices obtained by deleting the first row and column, or the first two rows and two columns, ultimately produces the desired form, apart from order. Since the summands can be rearranged at pleasure by simultaneous row and column permutations, the proof is complete.

9. Congruence over a field of characteristic $\neq 2$

When F is not of characteristic 2, this theorem can be simplified. For then it is easy to show that

$$\begin{bmatrix} 0 & 1 \\ 1 & 0 \end{bmatrix} \tilde{C} \operatorname{diag}(\tfrac{1}{2}, -\tfrac{1}{2})$$

and we get

Theorem IV.7. Let A be a symmetric matrix of rank r belonging to F_n, where F is not of characteristic 2. Then

$$A \tilde{C} \operatorname{diag}(d_1, d_2, \ldots, d_r) \dotplus 0$$

where $d_i \neq 0$, $1 \leq i \leq r$.

The exception is a true exception, for if F is of characteristic 2, $\begin{bmatrix} 0 & 1 \\ 1 & 0 \end{bmatrix}$ is not congruent to a diagonal matrix.

10. Congruence over an algebraically closed field of characteristic ≠ 2

The preceding result allows a complete solution in the case of an algebraically closed field of characteristic different from 2. We have

Theorem IV.8. Let F be an algebraically closed field of characteristic different from 2. Then two symmetric matrices of F_n are congruent if and only if they have the same rank.

Proof. Let A be any symmetric matrix of F_n of rank r. Then the preceding theorem implies that

$$A \tilde{C} \operatorname{diag}(d_1, d_2, \ldots, d_r) \dotplus 0,$$

$d_i \neq 0$, $1 \leq 1 \leq r$. Since F is algebraically closed, we can find elements $\beta_1, \beta_2, \ldots, \beta_r$ of F such that $\beta_i^2 d_i = 1$, $1 \leq i \leq r$. Put

$$U = \operatorname{diag}(\beta_1, \beta_2, \ldots, \beta_r) \dotplus I_{n-r}$$

Then U is a unit matrix of F_n and

$$U^{\mathrm{T}}\{\operatorname{diag}(d_1, d_2, \ldots, d_r) \dotplus 0\}U = I_r \dotplus 0$$

Thus every matrix of rank r is congruent to $I_r \dotplus 0$. This completes the proof.

11. Witt's theorem

The next result is basic when dealing with congruence over a field of characteristic different from 2.

Theorem IV.9 (Witt's cancellation theorem). Let F be a field of characteristic different from 2. Let A, B, C be symmetric matrices over F such that $A \dotplus B \tilde{C} A \dotplus C$. Then $B \tilde{C} C$.

Proof. By virtue of Theorem IV.7, we may assume that A is diagonal.

Thus it is clearly sufficient to prove the theorem for the case when A is the one element matrix (a). Suppose then that P is a unit matrix such that

$$\begin{bmatrix} a & 0 \\ 0 & B \end{bmatrix} = P^{\mathrm{T}} \begin{bmatrix} a & 0 \\ 0 & C \end{bmatrix} P$$

Partition P conformally, so that

$$P = \begin{bmatrix} p & x \\ y & Q \end{bmatrix}$$

where p is an element of F. Then

$$y^{\mathrm{T}}Cy = a(1 - p^2), \qquad y^{\mathrm{T}}CQ = -apx, \qquad Q^{\mathrm{T}}CQ = B - ax^{\mathrm{T}}x$$

We first modify Q to produce a congruence relationship between B and C. Put $W = Q - \theta yx$, where θ is an element of F to be determined later. Then a brief computation using the properties above shows that

$$W^{\mathrm{T}}CW = B - a(p\theta - \theta - 1)(p\theta + \theta - 1)x^{\mathrm{T}}x$$

Since F is not of characteristic 2, θ may be determined so that at least one of the equations

$$(p - 1)\theta - 1 = 0, \qquad (p + 1)\theta - 1 = 0$$

has a solution. For this choice of θ,

$$W^{\mathrm{T}}CW = B$$

The proof is not yet complete: W may be singular. Now B and C are certainly of the same rank [since $(a) \dotplus B$ and $(a) \dotplus C$ are of the same rank] so that we can write

$$B = U^{\mathrm{T}} \begin{bmatrix} D & 0 \\ 0 & 0 \end{bmatrix} U, \qquad C = V^{\mathrm{T}} \begin{bmatrix} E & 0 \\ 0 & 0 \end{bmatrix} V$$

where U, V are unit matrices and D, E are nonsingular $r \times r$ matrices. Thus it suffices to show that $\left[\begin{smallmatrix} D & 0 \\ 0 & 0 \end{smallmatrix}\right]$, $\left[\begin{smallmatrix} E & 0 \\ 0 & 0 \end{smallmatrix}\right]$ are congruent. Put $T = VWU^{-1}$. Then

$$T^{\mathrm{T}} \begin{bmatrix} E & 0 \\ 0 & 0 \end{bmatrix} T = \begin{bmatrix} D & 0 \\ 0 & 0 \end{bmatrix}$$

Put $T = \left[\begin{smallmatrix} T_1 & T_2 \\ T_3 & T_4 \end{smallmatrix}\right]$, where T_1 is $r \times r$. Then $T_1{}^{\mathrm{T}}ET_1 = D$. Since E and D are nonsingular, T_1 is nonsingular. But then $S = \left[\begin{smallmatrix} T_1 & 0 \\ 0 & I \end{smallmatrix}\right]$ is also nonsingular, and

$$S^{\mathrm{T}}\begin{bmatrix} E & 0 \\ 0 & 0 \end{bmatrix} S = \begin{bmatrix} D & 0 \\ 0 & 0 \end{bmatrix}$$

Hence B and C are congruent, and the proof is concluded.

12. Two lemmas on finite fields

Witt's theorem will find an application to the case when F is a finite field. But first we prove two lemmas.

Lemma IV.2. In a finite field F, every element is the sum of two squares.

Proof. Suppose that F has q elements. If F is of characteristic 2 then q is even, and so every element α of F is a square, since $\alpha = \alpha^q$. Suppose then that q is odd. Then there are precisely $(q+1)/2$ squares in F, consisting of 0 and g^{2k}, $0 \leq k \leq (q-3)/2$, where g is a generator of the multiplicative group of F.

Let S be the set of squares. Let α be any element of F. Consider the set $\{\alpha - x : x \in S\}$. This set consists of $(q+1)/2$ distinct elements of F, and so cannot be disjoint from S, since F has just q elements in all. It follows that elements x, y of S exist such that $\alpha - x = y$. Thus α is the sum of two squares.

Lemma IV.3. Let α be any nonzero element of the finite field F. Then

$$\alpha I_2 \mathrel{\tilde{C}} I_2$$

Proof. By the previous lemma, elements x, y of F exist such that $\alpha = x^2 + y^2$. Put $P = \left[\begin{smallmatrix} x & y \\ -y & x \end{smallmatrix}\right]$. Then $\det(P) = x^2 + y^2 = \alpha \neq 0$, so that P is a unit matrix of F_2. Also,

$$P^{\mathrm{T}} I_2 P = \begin{bmatrix} x & -y \\ y & x \end{bmatrix}\begin{bmatrix} x & y \\ -y & x \end{bmatrix} = (x^2 + y^2) I_2 = \alpha I_2$$

This concludes the proof.

13. Congruence over a field of finite characteristic

We can now prove the following theorem.

Theorem IV.10. Let F be a finite field of characteristic different from 2. Then every symmetric matrix of F_n of rank r is congruent to $I_r \dotplus 0$ or to $gI_1 \dotplus I_{r-1} \dotplus 0$, and these are not congruent. Thus the symmetric matrices of any given positive rank fall into precisely two congruence classes.

Proof. Let A belong to F_n, A symmetric of rank r. By Theorem IV.7,

$$A \tilde{C} \operatorname{diag}(\alpha_1, \alpha_2, \ldots, \alpha_r) \dotplus 0$$

where $\alpha_i \neq 0$, $1 \leq i \leq r$. We may write

$$\alpha_i = g^{e_i}\beta_i^2, \qquad 1 \leq i \leq r$$

where $e_i = 0$ or 1 and $\beta_i \in F$. Then if $D = \operatorname{diag}(\beta_1^{-1}, \beta_2^{-1}, \ldots, \beta_r^{-1}) \dotplus I_{n-r}$, D is a unit matrix of F_n and

$$D^T(\operatorname{diag}(\alpha_1, \alpha_2, \ldots, \alpha_r) \dotplus 0)D = \operatorname{diag}(g_1^{e_1}, g_2^{e_2}, \ldots, g_r^{e_r}) \dotplus 0$$

It follows that

$$A \tilde{C} gI_s \dotplus I_{r-s} \dotplus 0$$

for some s such that $0 \leq s \leq r$.

By repeated use of Lemma IV.3, we find that if s is even, then $gI_s \tilde{C} I_s$ and if s is odd, then

$$gI_s \tilde{C} gI_1 \dotplus I_{s-1}$$

Thus A is congruent to $I_r \dotplus 0$ or to $gI_1 \dotplus I_{r-1} \dotplus 0$.

Suppose now that

$$I_r \dotplus 0 \tilde{C} gI_1 \dotplus I_{r-1} \dotplus 0$$

Then Witt's cancellation theorem implies that $I_1 \tilde{C} gI_1$, making g a square, which it is not. This concludes the proof.

14. Two lemmas on finite fields of characteristic 2

We now treat the case of a finite field F of characteristic 2. Let

$$T = \begin{bmatrix} 0 & 1 \\ 1 & 0 \end{bmatrix}$$

and let qT denote the direct sum of q copies of T. We require two lemmas.

Lemma IV.4. Every matrix congruent to qT has all diagonal elements zero.

Proof. Let U be a unit matrix of F_{2q}, and consider $A = U^{\mathrm{T}}qTU$. Write $U = (U_{ij})$, $A = (A_{ij})$, where each U_{ij} and each A_{ij} belong to F_2. Then the diagonal elements of A arise from the diagonal blocks

$$A_{ii} = \sum_{k=1}^{q} U_{ki}^{\mathrm{T}} T U_{ki}$$

But for any matrix B of F_2,

$$B^{\mathrm{T}}TB = \det(B)T$$

Thus

$$A_{ii} = \left\{\sum_{k=1}^{q} \det(U_{ki})\right\} T$$

and hence has zero diagonal elements.

Lemma IV.5. Over F_3,

$$I_1 \dotplus T \tilde{C} I_3$$

Proof. Put

$$P = \begin{bmatrix} 1 & 1 & 1 \\ 0 & 1 & 1 \\ 1 & 0 & 1 \end{bmatrix}$$

Then P is a unit matrix of F_3 and

$$PI_3P^{\mathrm{T}} = PP^{\mathrm{T}} = I_1 \dotplus T$$

15. Congruence over finite fields of characteristic 2

We can now prove the analogue of Theorem IV.10 for finite fields of characteristic 2. We have

Theorem IV.11. Let F be a finite field of characteristic 2. Then every symmetric matrix of F_n of rank r is congruent to $I_r \dotplus 0$ (r odd), or to $I_r \dotplus 0$ or $(r/2)\ T \dotplus 0$ (r even), and these are not congruent.

Proof. Let A be any symmetric matrix of rank r of F_n. By Theorem IV.6 A is congruent to

$$\operatorname{diag}(\alpha_1, \alpha_2, \ldots, \alpha_p) \dotplus qT \dotplus 0$$

where $\alpha_i \neq 0$, $1 \leq i \leq p$, and $p + 2q = r$. Since every element of a finite field of characteristic 2 is a square, there are elements $\beta_1, \beta_2, \ldots, \beta_p$ of F such that $\beta_i^2\alpha_i = 1$, $1 \leq i \leq p$. Put $D = \operatorname{diag}(\beta_1, \beta_2, \ldots, \beta_p)$. Then

$$D^{\mathrm{T}} \operatorname{diag}(\alpha_1, \alpha_2, \ldots, \alpha_p)D = I_p$$

It follows that A is congruent to

$$I_p \dotplus qT \dotplus 0$$

Now if $p \neq 0$, repeated use of Lemma IV.5 implies that A is congruent to $I_{p+2q} \dotplus 0$; and if $p = 0$, A is congruent to $qT \dotplus 0$. We thus obtain the two types of the theorem. It only remains to verify that they are not congruent. But this is implied by Lemma IV.4, since any matrix congruent to $qT \dotplus 0$ must also have all diagonal elements equal to zero. This completes the proof.

16. Ordered fields

Our next result concerns ordered fields. A field F is *ordered* if there is a nonempty subset P, the elements of which are referred to as *positive*, such that P is closed with respect to addition and multiplication, and such that every element a of F satisfies just one of the following conditions:

a is positive
a is zero
$-a$ is positive (a is *negative*).

By $a > b$ (or $b < a$) we mean that $a - b$ is positive. Notice that an ordered field cannot be of finite characteristic.

17. Sylvester's law of inertia

The next result is fundamental to the study of congruence over ordered fields.

Theorem IV.12 (Sylvester's law of inertia). Let $D = \operatorname{diag}(d_1, d_2, \ldots, d_r) \dotplus 0$ and $E = \operatorname{diag}(e_1, e_2, \ldots, e_r) \dotplus 0$ be congruent matrices of rank r with elements in an ordered field F. Then the number of positive d_i is the same as the number of positive e_i.

Proof. By Witt's cancellation theorem, we have

$$\operatorname{diag}(d_1, d_2, \ldots, d_r)\ \tilde{C}\ \operatorname{diag}(e_1, e_2, \ldots, e_r)$$

Let $U = (u_{ij})$ be a unit matrix of F, such that

$$\operatorname{diag}(d_1, d_2, \ldots, d_r) = U^{\mathrm{T}}\operatorname{diag}(e_1, e_2, \ldots, e_r)U$$

Let x be an $r \times 1$ vector with entries in F and put $y = Ux$. Then

$$\sum_{i=1}^{r} d_i x_i^2 = \sum_{i=1}^{r} e_i y_i^2$$

where $x = [x_1, x_2, \ldots, x_r]^{\mathrm{T}}$, $y = [y_1, y_2, \ldots, y_r]$ We may assume that $d_1, d_2, \ldots, d_p$ are the positive elements among the d_i and that $e_1, e_2, \ldots, e_q$ are the positive elements among the e_i. Suppose that $p < q$. Choose

$$x_1 = x_2 = \cdots = x_p = 0$$

Then

$$y_i = \sum_{j=p+1}^{r} u_{ij} x_j, \qquad 1 \leq i \leq r$$

The equations

$$\sum_{j=p+1}^{r} u_{ij} x_j = 0, \qquad q + 1 \leq i \leq r$$

form a system of $r - q$ homogeneous equations in the $r - p$ unknowns $x_{p+1}, x_{p+2}, \ldots, x_r$; and since $p < q$, $r - p > r - q$, so that there are more unknowns than equations. Hence this system has a nontrivial solution in F, and so we can choose

$$x_1 = x_2 = \cdots = x_p = 0$$
$$x_{p+1}, x_{p+2}, \ldots, x_r \text{ not all } 0$$

such that

$$y_{q+1} = y_{q+2} = \cdots = y_r = 0$$

But then

$$e_1 y_1^2 + e_2 y_2^2 + \cdots + e_q y_q^2 - d_{p+1} x_{p+1}^2 - d_{p+2} x_{p+2}^2 - \cdots - d_r x_r^2 = 0$$

and

$$e_i > 0, \qquad 1 \leq i \leq q, \qquad -d_i > 0, \qquad p + 1 \leq i \leq r$$

Since the square of any nonzero element in an ordered field is positive, we arrive at a contradiction. Hence $p \geq q$; and similarly, $q \geq p$. Thus $p = q$ and the theorem is proved.

18. Congruence over the real field

We have thus found an invariant for symmetric matrices A over an ordered field F; namely, the number of positive terms in the diagonal form of A. We refer to this number as the *index* of A. Then the question of congruence over the real field is completely settled by the following theorem:

Theorem IV.13. Two real symmetric $n \times n$ matrices A, B are congruent if and only if they have the same rank and the same index.

Proof. In fact, if p is the index and r is the rank, then each of A, B is congruent to

$$I_p \dot{+} -I_{r-p} \dot{+} 0$$

and in view of Sylvester's law of inertia, this form is unique.

19. Positive definite matrices

A symmetric $n \times n$ matrix A with elements in an ordered field F is *positive definite* if its rank and index are both n. If F is the field of reals, then Theorem IV.13 shows that A is positive definite if and only if $A \tilde{C} I_n$. Thus A is positive definite if and only if there is a real, nonsingular matrix P such that $A = P^T P$.

20. Congruence over the rational field

We now come to congruence over the rational field Q. This subject constitutes a landmark in the development of number theory, beginning with the work of Minkowski in the late nineteenth century, and culminating in the work of Hasse. Since this topic properly belongs to number theory, and since excellent accounts of it are available in many places (for example, the book by Borevich and Shafarevich [3]) we will not discuss it here in any detail, but content ourselves with the statement of the principal theorem of the subject.

Theorem IV.14 (Hasse–Minkowski theorem). Let A, B be $n \times n$ symmetric matrices over Q. Then A and B are congruent over Q if and only if they have the same rank and index, and are congruent over the p-adic number field $F(p)$ for every prime p.

The content of this theorem is that a "global" result is deduced from infinitely many "local" ones, and the method so introduced into number theory has been epoch making.

21. Congruence over Z

We now leave fields and consider congruence over Z. We will develop this subject at least to the point where the finiteness of the number of congruence clases can be proved for matrices of given determinant which do not represent 0 properly, and then leave it. We will return to certain aspects of the subject in Chapter XI.

22. The arithmetic minimum

Let $A = (a_{ij})$ be a symmetric matrix belonging to Z_n which does not represent 0 properly. Let V_n be the set of $n \times 1$ vectors over Z. The set

$$q(A) = \{x^{\mathrm{T}}Ax : x \in V_n\}$$

is a set of integers, and consequently there is a nonzero y in V_n such that $|y^{\mathrm{T}}Ay|$ is least, and hence positive, since $x^{\mathrm{T}}Ax = 0$ for x in V_n implies that $x = 0$. If both $y^{\mathrm{T}}Ay$ and $-y^{\mathrm{T}}Ay = z^{\mathrm{T}}Az$ occur, we resolve the ambiguity by choosing the positive value. We write $m(A) = y^{\mathrm{T}}Ay$, and refer to it as the *arithmetic minimum* of A. Clearly, the coordinates of the minimizing vector y must be relatively prime, and hence Lemma IV.1 implies that $m(A)$ occurs as the (1, 1) element of some matrix congruent to A.

23. The case $n = 2$

We first consider $n = 2$.

Let $A = \left[\begin{smallmatrix} a & b \\ b & c \end{smallmatrix}\right]$ be any symmetric matrix of Z_2 which does not represent 0 properly. Put $d = \det(A) = ac - b^2$. Then after a suitable congruence transformation has been performed, we may assume that $a = m(A)$.

Now consider the congruence transformation

$$\begin{bmatrix} 1 & 0 \\ x & 1 \end{bmatrix} \begin{bmatrix} a & b \\ b & c \end{bmatrix} \begin{bmatrix} 1 & x \\ 0 & 1 \end{bmatrix} = \begin{bmatrix} a & * \\ ax + b & * \end{bmatrix}$$

where x belongs to Z. Since $a \neq 0$, we may choose x so that $|ax + b| \leq \frac{1}{2}|a|$. Furthermore, a is unchanged. Thus we may assume that after a suitable congruence transformation has been performed, the entries of A satisfy

$$(5) \qquad 2|b| \leq |a| \leq |c|$$

the last inequality holding because c is a value assumed by A (in fact for the vector $[0, 1]^T$). But then

$$a^2 \leq |ac| = |b^2 + d| \leq \tfrac{1}{4}a^2 + |d|, \qquad |a| \leq (2/\sqrt{3})|d|^{1/2}$$

Thus we have proved

Theorem IV.15. Let A be a symmetric matrix of Z_2 of determinant d which does not represent 0 properly. Then

$$(6) \qquad |m(A)| \leq (2/\sqrt{3})|d|^{1/2}$$

24. The finiteness of the class number for $n = 2$

Another consequence of our analysis is

Theorem IV.16. There are only finitely many congruence classes of symmetric matrices of Z_2 of given determinant d which do not represent 0 properly.

Proof. We have shown that every such matrix $A = \left[\begin{smallmatrix} a & b \\ b & c \end{smallmatrix}\right]$ is congruent to one for which the inequalities (5) hold. There are only finitely many choices for a (in fact, at most $2[(2/\sqrt{3})|d|^{1/2}]$) and only finitely many choices for b (at most $2|a| + 1$); and since c is determined once a and b are chosen from the relationship $ac - b^2 = d$, there are only finitely many possibilities in all. This concludes the proof.

25. The general case: Hermite's method

Having disposed of $n = 2$, we treat the general case by a simple inductive procedure due to Hermite which amounts to completing the square.

Let $A = (a_{ij})$ be any symmetric matrix of Z_n which does not represent 0 properly. Then no diagonal element of A, or of any matrix congruent to A, is zero. Put

$$U = \begin{bmatrix} 1 & a_{12}/a_{11} & a_{13}/a_{11} & \cdots & a_{1n}/a_{11} \\ 0 & 1 & 0 & \cdots & 0 \\ 0 & 0 & 1 & \cdots & 0 \\ & \cdot & & & \\ & \cdot & & & \\ & \cdot & & & \\ 0 & 0 & 0 & \cdots & 1 \end{bmatrix}$$

The matrix U is not in general a matrix of Z_n, but is certainly a matrix over Q of determinant 1. Then a short computation shows that

$$A = U^{\mathrm{T}} \begin{bmatrix} a_{11} & 0 \\ 0 & (1/a_{11})B \end{bmatrix} U \tag{7}$$

where B belongs to Z_{n-1}, and has determinant

$$\det(B) = a_{11}^{n-2} \det(A)$$

In terms of quadratic forms, let x be any vector of V_n and put $y = Ux$, so that

$$\begin{aligned} y_1 &= x_1 + \frac{a_{12}}{a_{11}} x_2 + \frac{a_{13}}{a_{11}} x_3 + \cdots + \frac{a_{1n}}{a_{11}} x_n, \\ y_2 &= x_2, \\ &\cdot \\ &\cdot \\ &\cdot \\ y_n &= x_n \end{aligned}$$

Then if z is the vector $[y_2, y_3, \ldots, y_n]^{\mathrm{T}} = [x_2, x_3, \ldots, x_n]^{\mathrm{T}}$,

$$x^{\mathrm{T}}Ax = a_{11}y_1{}^2 + (1/a_{11})z^{\mathrm{T}}Bz$$

We may use (7) to derive a bound for $m(A)$. We first remark that B is also a matrix which does not represent 0 properly. For if z is a nonzero vector of V_{n-1} such that $z^{\mathrm{T}}Bz = 0$, then the same is true for $a_{11}z$. But then an integral value for x_1 may be determined which makes y_1 zero, implying that A represents 0 properly, which it does not.

Suppose now that a suitable congruence has been performed so that $a_{11} = m(A)$. Assume that for all $(n-1) \times (n-1)$ symmetric matrices

B which do not represent 0 properly, the inequality

$$|m(B)| \leq (\tfrac{4}{3})^{(n-2)/2} |\det(B)|^{1/(n-1)}$$

is valid. Choose z so that $z^T Bz = m(B)$. Now choose x_1 so that $|y_1| \leq \frac{1}{2}$. Then

$$\begin{aligned} |x^T Ax| &\leq \tfrac{1}{4}|a_{11}| + (\tfrac{4}{3})^{(n-2)/2} \frac{1}{|a_{11}|} |a_{11}^{n-2} \det(A)|^{1/(n-1)} \\ &= \tfrac{1}{4}|a_{11}| + (\tfrac{4}{3})^{(n-2)/2} |a_{11}|^{-1/(n-1)} |\det(A)|^{1/(n-1)} \end{aligned}$$

Since a_{11} is the arithmetic minimum of A, we have $|a_{11}| \leq |x^T Ax|$, which implies that

$$\begin{aligned} \tfrac{3}{4}|a_{11}| &\leq (\tfrac{4}{3})^{(n-2)/2} |a_{11}|^{-1/(n-1)} |\det(A)|^{1/(n-1)}, \\ |a_{11}|^{n/(n-1)} &\leq (\tfrac{4}{3})^{n/2} |\det(A)|^{1/(n-1)}, \\ |a_{11}| &\leq (\tfrac{4}{3})^{(n-1)/2} |\det(A)|^{1/n} \end{aligned}$$

Thus we have proved by induction that the following is valid:

Theorem IV.17. Let A be a symmetric matrix of Z_n which does not represent 0 properly. Then

$$|m(A)| \leq (\tfrac{4}{3})^{(n-1)/2} |\det(A)|^{1/n}$$

26. The finiteness of the class number for any *n*

We now consider the problem of the finiteness of the classes. Theorem IV.16 settles the question for $n = 2$, and we proceed by induction on n.

Suppose that we have shown that there are only finitely many congruence classes of symmetric matrices of Z_{n-1} with given determinant which do not represent 0 properly. Let $A = (a_{ij})$ be any such matrix of Z_n with given determinant d. We may assume that $a_{11} = m(A)$, after a suitable congruence transformation has been performed.

We now refer to formula (7). Write

$$U = \begin{bmatrix} 1 & \alpha \\ 0 & I_{n-1} \end{bmatrix}$$

where α is the vector

$$\alpha = [a_{12}/a_{11}, a_{13}/a_{11}, \ldots, a_{1n}/a_{11}]$$

Then

$$A = \begin{bmatrix} 1 & 0 \\ \alpha^{\mathrm{T}} & I_{n-1} \end{bmatrix} \begin{bmatrix} a_{11} & 0 \\ 0 & (1/a_{11})B \end{bmatrix} \begin{bmatrix} 1 & \alpha \\ 0 & I_{n-1} \end{bmatrix}$$

Since the possible values for a_{11} are limited by Theorem IV.17, and since $\det(B) = a_{11}^{n-2}d$, the possible values for the determinant of B are also limited. By the induction hypothesis, B is congruent to finitely many matrices, say $B_1, B_2, \ldots, B_p$. Thus for some unit matrix V of Z_{n-1} and for some k such that $1 \leq k \leq$ p, $B = V^{\mathrm{T}}B_kV$. It follows that

$$A = \begin{bmatrix} 1 & 0 \\ \alpha^{\mathrm{T}} & V^{\mathrm{T}} \end{bmatrix} \begin{bmatrix} a_{11} & 0 \\ 0 & (1/a_{11})B_k \end{bmatrix} \begin{bmatrix} 1 & \alpha \\ 0 & V \end{bmatrix}$$

Put $W = [\begin{smallmatrix} 1 & 0 \\ 0 & V^{-1} \end{smallmatrix}]$. Then W belongs to Z_n' and

$$A_1 = W^{\mathrm{T}}AW = \begin{bmatrix} 1 & 0 \\ \beta^{\mathrm{T}} & I_{n-1} \end{bmatrix} \begin{bmatrix} a_{11} & 0 \\ 0 & (1/a_{11})B_k \end{bmatrix} \begin{bmatrix} 1 & \beta \\ 0 & I_{n-1} \end{bmatrix}$$

where $\beta = \alpha V^{-1}$.

Now choose u in V_{n-1} so that the vector $u + \beta = y$ has each entry of absolute value at most $\frac{1}{2}$. Put

$$X = \begin{bmatrix} 1 & u \\ 0 & I_{n-1} \end{bmatrix}$$

Then

$$A_2 = X^{\mathrm{T}}A_1X = \begin{bmatrix} 1 & 0 \\ y^{\mathrm{T}} & I_{n-1} \end{bmatrix} \begin{bmatrix} a_{11} & 0 \\ 0 & (1/a_{11})B_k \end{bmatrix} \begin{bmatrix} 1 & y \\ 0 & I_{n-1} \end{bmatrix}$$

It is clear that there are only finitely many possibilities for A_2. Hence the induction is complete, and we have proved

Theorem IV.18. There are only finitely many congruence classes of symmetric matrices of Z_n of given determinant which do not represent 0 properly.

It is possible to arrange the preceding proof to produce an actual bound for the number of classes in terms of n and the determinant. There is little value in doing this, however, since other methods will produce much better bounds, and some of these will be discussed later.

A particularly important class of matrices which do not represent 0 properly is the class of positive definite symmetric matrices, to which the previous considerations apply.

EXERCISES AND PROBLEMS

1. Let A, B be positive definite symmetric matrices, $A \in Z_m$, $B \in Z_n$. Then

$$m(A \dotplus B) = \min\{m(A), m(B)\}$$
$$m(A \otimes B) \leq m(A)\, m(B)$$

2. Explain the relevance of the congruence of Lemma IV.5 to Witt's theorem.

3. Prove that if A is a symmetric matrix of Z_2 which represents 0 properly, then $\det(A) = -\Delta^2$, and $A \mathrel{\tilde{C}} \left[\begin{smallmatrix} 0 & \Delta \\ \Delta & c \end{smallmatrix}\right]$, where c may be taken modulo 2Δ. Are any of these congruent?

4. Suppose that A is a symmetric matrix of Z_n which represents 1. Prove that $A \mathrel{\tilde{C}} I_1 \dotplus B$, where $B \in Z_{n-1}$.

5. Prove that if $n \leq 5$, then every symmetric positive definite matrix of determinant 1 which belongs to Z_n is of the form $A^{\mathrm{T}}A$, where $A \in Z_n$.

Hint: Use Theorem IV.17 and the fact that $(\frac{4}{3})^{(n-1)/2} < 2$ for $n \leq 5$. The result is actually true for $n \leq 7$.

6. Show that if p is a prime of the form $4n + 1$, then p is the sum of two squares.

Hint: First show that there are integers a, m such that $a^2 + 1 = mp$, and consider the matrix $\left[\begin{smallmatrix} p & a \\ a & m \end{smallmatrix}\right]$.

7. Show that if p is a prime of the form $8n + 3$, then p is the sum of three squares.

Hint: First show that there are integers a, m such that $2a^2 + 1 = (2m - 1)p$, and consider the matrix

$$\begin{bmatrix} p & a & 0 \\ a & m & 1 \\ 0 & 1 & 2 \end{bmatrix}$$

8. Show that if p is any prime, then p is the sum of four squares.

Hint: First show that there are integers a, b, m, such that $a^2 + b^2 + 1 = mp$, and consider the matrix

$$\begin{bmatrix} p & 0 & a & b \\ 0 & p & -b & a \\ a & -b & m & 0 \\ b & a & 0 & m \end{bmatrix}$$

9. If P is the set of positive elements of an ordered field F, then P is a subgroup of index 2 of the multiplicative group of F which is closed with respect to addition. Is the existence of such a subgroup sufficient to guarantee that F is an ordered field?

Chapter V

Combined Similarity and Congruence

We now study a much more restrictive kind of equivalence.

1. Orthogonal equivalence

Let A, B be elements of R_n. We say that A is *orthogonally equivalent* to B if there is a unit matrix U of R_n such that $A = U^{\mathrm{T}}BU$, and U satisfies

$$UU^{\mathrm{T}} = I$$

A matrix of this type will be referred to as an *orthogonal* matrix. The set of orthogonal matrices U constitutes a subgroup O_n of the group of unit matrices R_n'. O_n is certainly not a trivial group, for the generalized permutation matrices all belong to O_n and there are $2^n n!$ of these (a generalized permutation matrix is one with just one nonzero entry, ± 1, in each row and column). However, it is quite conceivable that there are no others. This is so, for example, for the ring Z_n.

2. Orthogonal equivalence over a field of characteristic $\neq 2$

The theory of orthogonal equivalence (and its natural generalization, unitary equivalence, which will be defined later) has been worked out extensively for the real and complex fields. There is also a significant result for any field of characteristic different from 2, due to Hall and Ryser [14]. Apart from these topics, most work has been concerned with the structure of O_n for appropriate underlying rings, and not so much with canonical forms under orthogonal transformations.

The principal result of Hall and Ryser is a theorem analogous to Theorem II.1 and states that if $x = [x_1, x_2, \ldots, x_n]$, $y = [y_1, y_2, \ldots, y_n]$ are vectors over a field F of characteristic different from 2 such that

$$x_1^2 + x_2^2 + \cdots + x_n^2 = y_1^2 + y_2^2 + \cdots + y_n^2 = a \neq 0$$

then there is an orthogonal matrix U of F_n such that $y = xU$. In particular, if $a = 1$ and y is chosen as the vector $[1, 0, \ldots, 0]$, the theorem says that any vector of "unit length" may be completed to an orthogonal matrix. In this form the result is familiar for the real field, but not at all obvious even for the rationals, say.

3. A lemma of Hall and Ryser

We first prove

Theorem V.1. Let F be a field of characteristic different from 2. Let A belong to F_n, and suppose that A is nonsingular and that $AA^{\mathrm{T}} = D_1 \dotplus D_2$, where D_1 belongs to F_r and D_2 belongs to F_s, so that $r + s = n$. Let X be an arbitrary $r \times n$ matrix over F such that $XX^{\mathrm{T}} = D_1$. Then there is a matrix Z of F_n having X as its first r rows such that $ZZ^{\mathrm{T}} = D_1 \dotplus D_2$.

Proof. Since D_1 is of rank r and $XX^{\mathrm{T}} = D_1$, X is of rank r. Thus the null space of X is of dimension $n - r = s$. Let W be an $s \times n$ matrix whose rows are a basis for the null space of X, so that W is of rank s. Then if

$$Y = \begin{bmatrix} X \\ W \end{bmatrix}$$

$YY^{\mathrm{T}} = D_1 \dotplus D_3$, where $D_3 = WW^{\mathrm{T}}$.

We first show that Y is nonsingular. Let C be a nonsingular matrix of F_r such that

$$CD_1C^{\mathrm{T}} = D = \operatorname{diag}(d_1, d_2, \ldots, d_r)$$

so that $d_k \neq 0$, $1 \leq k \leq r$. Then

$$\begin{bmatrix} CX \\ W \end{bmatrix} [X^{\mathrm{T}}C^{\mathrm{T}}, W^{\mathrm{T}}] = \begin{bmatrix} D & 0 \\ 0 & D_3 \end{bmatrix}$$

Denote the row vectors of $T = [{}^{CX}_{W}]$ by $\alpha_1, \alpha_2, \ldots, \alpha_r, \beta_1, \beta_2, \ldots, \beta_s$. Then T is nonsingular. For if

$$\sum_{i=1}^{r} a_i\alpha_i + \sum_{j=1}^{s} b_i\beta_j = 0, \qquad a_i, b_j \in F$$

then the facts that $\alpha_i\alpha_k^{\mathrm{T}} = d_i\delta_{ik}$ with $d_i \neq 0$, and $\alpha_k\beta_j^{\mathrm{T}} = 0$, imply that $a_1 = a_2 = \cdots = a_r = 0$. But then the linear independence of $\beta_1, \beta_2, \ldots, \beta_s$ implies that $b_1 = b_2 = \cdots = b_s = 0$ as well. Hence T is nonsingular, and so Y is also nonsingular.

Now

$$D_1 \dotplus D_2 = AA^{\mathrm{T}} \tilde{C}\, I, \qquad D_1 \dotplus D_3 = YY^{\mathrm{T}} \tilde{C}\, I$$

so that

$$D_1 + D_2 \tilde{C}\, D_1 \dotplus D_3$$

Witt's cancellation theorem thus implies that

$$D_2 \tilde{C}\, D_3$$

Put $D_2 = ED_3E^{\mathrm{T}}$, E nonsingular, and put $Z_2 = [{}^{X}_{EW}]$. Then

$$ZZ^{\mathrm{T}} = \begin{bmatrix} D_1 & 0 \\ 0 & EWW^{\mathrm{T}}E^{\mathrm{T}} \end{bmatrix} = \begin{bmatrix} D_1 & 0 \\ 0 & D_2 \end{bmatrix}$$

This concludes the proof.

4. The theorem of Hall and Ryser

From this theorem we deduce

Theorem V.2. Let $x = [x_1, x_2, \ldots, x_n]$, $y = [y_1, y_2, \ldots, y_n]$ be vectors over F such that

$$x_1^2 + x_2^2 + \cdots + x_n^2 = y_1^2 + y_2^2 + \cdots + y_n^2 = a \neq 0$$

Then there is an orthogonal matrix U of F_n such that $y = xU$.

Proof. As in the first part of the proof of the preceding theorem, there is an $(n-1) \times n$ matrix W of rank $n-1$ such that the matrix $X = [{x \atop W}]$ satisfies

$$XX^{\mathrm{T}} = \begin{bmatrix} a & 0 \\ 0 & D \end{bmatrix}, \qquad D = WW^{\mathrm{T}}$$

Furthermore X is nonsingular, since $a \neq 0$. By the preceding theorem, there is a matrix $Y = [{y \atop K}]$ such that $YY^{\mathrm{T}} = [{a \atop 0}\ {0 \atop D}]$. Then the matrix $U = X^{-1}Y$ is orthogonal; and since

$$[1, 0, \ldots, 0]X = x, \qquad [1, 0, \ldots, 0]Y = y, \qquad y = xU$$

this completes the proof.

As was mentioned earlier, this theorem has the corollary

Corollary V.1. Let F be any field of characteristic different from 2. Let $x = [x_1, x_2, \ldots, x_n]$ be any vector over F such that

$$x_1^2 + x_2^2 + \cdots + x_n^2 = 1.$$

Then there is an orthogonal matrix of F_n having x as any prescribed row.

5. The complex field and unitary equivalence

We go on now to the classical situation of the complex field C. Because of the presence of the involutory automorphism $\alpha \longrightarrow \bar{\alpha}$, $\alpha \in C$, we introduce the notion of *unitary equivalence*, and discuss this fully. Orthogonal equivalence over the reals becomes a special case.

Suppose that A, B belong to C_n. We say that A is *unitarily equivalent* to B if there is a matrix U of C_n such that

$$A = \bar{U}^{\mathrm{T}}BU,$$

and U satisfies

$$\bar{U}^{\mathrm{T}}U = I \tag{1}$$

For brevity, we put $\bar{U}^{\mathrm{T}} = U^*$. Thus if $U = (u_{ij})$, then $U^* = (\bar{u}_{ji})$. A matrix satisfying (1) is referred to as a *unitary* matrix. A real unitary matrix is thus an orthogonal matrix.

6. Inner products

The situation is best discussed in terms of the underlying vector space E_n, the set of all $1 \times n$ vectors over C, known as complex, euclidean, n-dimensional space. E_n admits the inner product

$$(x, y) = x\bar{y}^{\mathrm{T}} = xy^*, \qquad x, y \in E_n$$

The inner product is scalar valued, and for all x, y, z of E_n and all a, b of C,

$$(x, y) = \overline{(y, x)}, \tag{2}$$

$$(ax + by, z) = a(x, z) + b(y, z), \tag{3}$$

$$(x, x) \geq 0; \qquad (x, x) = 0 \text{ if and only if } x = 0 \tag{4}$$

7. Length

This permits us to define the *length* of a vector x of E_n as the positive square root of (x, x):

$$\|x\| = (x, x)^{1/2}$$

Then (2), (3), (4) imply the *Schwarz* inequality:

$$|(x, y)| \leq \|x\| \cdot \|y\|, \qquad x, y \in E_n \tag{5}$$

Equality holds in (5) if and only if x and y are linearly dependent over C.

We say that x and y are *orthogonal* if $(x, y) = 0$, $x, y \in E_n$. A set of mutually orthogonal vectors of E_n will be called an *orthogonal* set.

8. The Gram–Schmidt process

Suppose that

$$x_1, x_2, \ldots, x_n \tag{6}$$

is a basis of E_n. We shall describe a process (the Gram–Schmidt orthogonalization process) for replacing the vectors of (6), one at a time, by a new set of vectors which also form a basis and in addition are an orthogonal set. We make the initial choice

$$y_1 = x_1$$

Suppose now that vectors $y_1, y_2, \ldots, y_{k-1}$, where $2 \leq k \leq n$, have been determined so that

$$\begin{aligned} y_1 &= x_1, \\ y_2 &= x_2 - a_{12} y_1, \\ &\vdots \\ y_{k-1} &= x_{k-1} - a_{1,k-1} y_1 - a_{2,k-1} y_2 - \cdots - a_{k-2,k-1} y_{k-2}, \end{aligned}$$

and $(y_i, y_j) = 0$, $1 \leq i, j \leq k-1$, $i \neq j$. Set

$$y_k = x_k - a_{1k} y_1 - a_{2k} y_2 - \cdots - a_{k-1,k} y_{k-1}$$

Suppose that $1 \leq i \leq k-1$. Then by the orthogonality,

$$(y_k, y_i) = (x_k, y_i) - a_{ik}(y_i, y_i)$$

Hence to make $(y_k, y_i) = 0$ it is necessary and sufficient to choose

$$a_{ik} = (x_k, y_i)/(y_i, y_i), \qquad 1 \leq i \leq k-1 \tag{7}$$

This choice is possible since $y_i \neq 0$, and so $(y_i, y_i) > 0$. We obtain therefore

Theorem V.3. Suppose that (6) is a basis of E_n. Define the vectors $y_1, y_2, \ldots, y_n$ recursively by

$$\begin{aligned} y_1 &= x_1, \\ y_k &= x_k - \sum_{i=1}^{k-1} a_{ik} y_i, \qquad 2 \leq k \leq n \end{aligned}$$

where a_{ik} is given by (7). Then the vectors $y_1, y_2, \ldots, y_n$ are an orthogonal basis of E_n.

The Gram–Schmidt process is actually valid for any field with no *isotropic* vectors (nonzero vectors of zero length) and hence for any ordered field, since all operations are field operations, and the only critical point is the nonvanishing of (x, x) for any nonzero x.

9. Normalized vectors and orthonormal bases

We shall say that a vector x of E_n is *normalized* if it is of unit length. If we define

$$z_i = y_i/\|y_i\|, \qquad 1 \leq i \leq n$$

then the vectors $z_1, z_2, \ldots, z_n$ are an orthogonal basis of E_n in which each basis element is of unit length. Such a basis is called an *orthonormal basis*.

10. Completion to a unitary matrix

An important consequence of Theorem V.3 is

Theorem V.4. Suppose that x_1 is a normalized vector of E_n. Then vectors $z_2, z_3, \ldots, z_n$ of E_n exist such that $x_1, z_2, \ldots, z_n$ is an orthonormal basis of E_n.

Proof. First extend x_1 to a basis of E_n, $x_1, x_2, \ldots, x_n$. Then apply the Gram–Schmidt process to determine an orthogonal basis $x_1, y_2, \ldots, y_n$. Finally, normalize to produce the orthonormal basis $z_1, z_2, \ldots, z_n$; and since x_1 is normalized, $z_1 = x_1$.

We note that the process may be applied to vectors in real euclidean n-dimensional space, and will produce vectors in real euclidean n-dimensional space.

If $z_1, z_2, \ldots, z_n$ is any orthonormal basis of E_n, then the matrix

$$U = \begin{bmatrix} z_1 \\ z_2 \\ \cdot \\ \cdot \\ \cdot \\ z_n \end{bmatrix}$$

satisfies

$$UU^* = (z_i z_j^*) = ((z_i, z_j)) = (\delta_{ij})$$

and consequently is unitary. Thus a restatement of Theorem V.4 is

Theorem V.5. Let x be any normalized vector of E_n. Then there is a unitary matrix of C_n having x as its first row.

Naturally, the same result holds for column vectors, or for vectors in any prescribed row or column position.

11. Schur's theorem on unitary equivalence

We can now state our first fundamental result on unitary equivalence:

Theorem V.6 (Schur). Let A be any matrix of C_n. Then A is unitarily equivalent to a triangular matrix.

Proof. Let λ_1 be an eigenvalue of A, and x_1 a corresponding eigenvector, which we may take to be normalized. Let $V = [x_1, x_2, \ldots, x_n]$ be a unitary matrix of C_n having x_1 as its first column. Then the first column of V^*AV is

$$\begin{bmatrix} x_1{}^*Ax_1 \\ x_2{}^*Ax_1 \\ \cdot \\ \cdot \\ \cdot \\ x_n{}^*Ax_1 \end{bmatrix} = \begin{bmatrix} \lambda_1 x_1{}^*x_1 \\ \lambda_1 x_2{}^*x_1 \\ \cdot \\ \cdot \\ \cdot \\ \lambda_1 x_n{}^*x_1 \end{bmatrix} = \begin{bmatrix} \lambda_1 \\ 0 \\ \cdot \\ \cdot \\ \cdot \\ 0 \end{bmatrix}$$

Repeating the process with the principal submatrix obtained by deleting the first row and column of V^*AV (which is equivalent to operating with unitary matrices of the form $I_1 \dotplus U$, $U \in C_{n-1}$) we ultimately obtain the desired form.

Notice that if A is a real matrix with real eigenvalues, then in fact $A = U^T T U$, where U is a real orthogonal matrix and T a real triangular matrix.

12. Normal matrices and unitary equivalence to a diagonal matrix

The next question we take up is when unitary equivalence to a diagonal matrix is possible. Suppose that $A = U^*DU$, where U is unitary and D diagonal. Then $AA^* = U^*DD^*U = U^*D^*DU = A^*A$, so that A commutes with A^*. Conversely, suppose that A commutes with A^*. Put $A = U^*TU$, where U is unitary and T is triangular. Then T also commutes with T^*. But this easily implies that T is in fact diagonal. If we say that A is *normal* if $AA^* = A^*A$, then this discussion shows

Theorem V.7. *A matrix A of C_n is unitarily equivalent to a diagonal matrix if and only if it is normal.*

Some classes of normal matrices are

hermitian matrices ($A^* = A$),
skew-hermitian matrices ($A^* = -A$),
real symmetric matrices,
real skew-symmetric matrices,
unitary matrices ($AA^* = I$),
real orthogonal matrices.

13. Miscellaneous results

The classical literature abounds with results on unitary and orthogonal equivalence, but we will prove no more here. The ones we have proved are basic, and can be used to show such results as the following:

Any complex matrix A may be written as $A = UH$, where U is unitary and H is hermitian.

Any complex matrix A is of the form $A = UDV$, where U, V are unitary and D is diagonal.

A real symmetric matrix is of the form $U^T DU$, where U is a real orthogonal matrix and D a real diagonal matrix.

Chapter VI

The Geometry of Numbers

This chapter provides a complete, self-contained introduction to the Minkowski geometry of numbers [34]. Its principal purpose is to make use of the powerful methods of the subject in deriving sharp bounds for the artithmetic minimum discussed previously. In addition, other interesting connections with number theory are pointed out and certain classical results are derived.

1. Lattices

Let E_n denote real n-dimensional euclidean space, consisting of all $n \times 1$ vectors (also referred to as points) with real coordinates. Let $\alpha_1, \alpha_2, \ldots, \alpha_n$ be any n linearly independent vectors of E_n. The totality L of vectors

$$g_1\alpha_1 + g_2\alpha_2 + \cdots + g_n\alpha_n$$

where $g_1, g_2, \ldots, g_n$ run over all integers, is called a *lattice* and will be denoted by

$$L = \{\alpha_1, \alpha_2, \ldots, \alpha_n\}$$

The lattice consisting of all $n \times 1$ vectors with integral coordinates will be denoted by G, and corresponds to the choice $\alpha_k = \delta_k$, the kth unit vector.

2. Lattice bases

A set of vectors β_k such that

$$L = \{\beta_1, \beta_2, \ldots, \beta_n\}$$

is called a *lattice basis* for L. Thus the vectors α_k are a lattice basis for L. Since each α_k is an integral linear combination of $\beta_1, \beta_2, \ldots, \beta_n$ and each β_k is an integral linear combination of $\alpha_1, \alpha_2, \ldots, \alpha_n$, integral $n \times n$ matrices M and N must exist such that

$$B = AM, \qquad A = BN$$

where A is the matrix $[\alpha_1, \alpha_2, \ldots, \alpha_n]$ and B the matrix $[\beta_1, \beta_2, \ldots, \beta_n]$. It follows that

$$A = AMN$$

and since A is nonsingular, $MN = I$. This implies that $\det(M) = \det(N) = \pm 1$. Conversely, if M is any $n \times n$ integral matrix of determinant ± 1 and

$$[\beta_1, \beta_2, \ldots, \beta_n] = [\alpha_1, \alpha_2, \ldots, \alpha_n]M$$

then the vectors β_k are a lattice basis of L, since M^{-1} is also an integral matrix.

3. The determinant of a lattice

The quantity

$$\Delta = |\det[\alpha_1, \alpha_2, \ldots, \alpha_n]| \tag{1}$$

is an invariant of L and is called the *determinant of the lattice L*. Δ can be interpreted geometrically as the volume of the parallelepiped based on the vectors $\overrightarrow{0\alpha_k}$, where 0 is the zero vector.

4. A geometrical consequence

An interesting geometrical conclusion can be drawn from this fact. Let $\xi_1, \xi_2, \ldots, \xi_n$ be n linearly independent vectors of G, and let P

be the parallelepiped based on the vectors $\overrightarrow{0\xi_k}$. Then P has volume 1 if and only if the ξ_k are a lattice basis for G. But also, the vectors ξ_k are a lattice basis if and only if P meets G in precisely its vertices. Hence we can conclude that *a parallelepiped whose vertices belong to G has volume* 1 *if and only if it meets G in precisely its vertices.*

5. Formal properties of lattices

The formal properties of lattices are very simple. A lattice is just the direct sum of n infinite cyclic groups, so that it is a free abelian group of rank n. Any subset of a lattice which is a module and which contains n linearly independent points is clearly also a lattice. The following theorem will be proved later:

Theorem VI.1. Let L be a subset of E_n. Then L is a lattice if and only if

(2) L is a module.
(3) L contains n linearly independent points.
(4) There is some sphere centered at 0 which meets L in only finitely many points.

6. Linear transformations

Let

(5) $$y = Px, \qquad \det(P) \neq 0$$

be a real linear transformation of E_n onto itself. We shall be concerned with the effect of P on various subsets of E_n; for example, convex subsets.

7. Convexity and central symmetry

A subset K of E_n is said to be *convex* if

(6) $x \in K, y \in K$ implies that $ax + by \in K$ for all $a, b \geq 0$ such that $a + b = 1$.

A subset K of E_n is said to have *center* 0 if

(7) $x \in K$ implies that $-x \in K$.

8. Preservation of convexity and central symmetry under linear transformations

Theorem VI.2. Let the transformation (5) map K onto K'. Then

(8) K is convex if and only if K' is convex.
(9) K has center 0 if and only if K' has center 0.

Proof. Since P is nonsingular, it is sufficient to prove the implications in one direction only. Suppose that K is convex. Let x', y' be points of K' and let $a, b \geq 0$, $a + b = 1$. Then (unique) points $x, y \in K$ exist such that $x' = Px$, $y' = Py$. Thus $ax + by \in K$, and so

$$ax' + by' = P(ax + by) \in K'$$

Hence K' is convex.

Suppose that K has center 0. Let $x' \in K'$. Then a (unique) point $x \in K$ exists such that $x' = Px$. Thus $-x \in K$, $-x' = P(-x) \in K'$. Hence K' has center 0. This completes the proof.

9. Volume under linear transformations

Suppose that K has a volume, and let $V(K)$ denote the volume of K. Then K' has a volume $V(K')$, and we have

Theorem VI.3. The volumes of K and K' satisfy

(10) $$V(K') = |\det(P)|\, V(K)$$

Proof. $\det(P)$ is just the Jacobian of the transformation (5).

If S is any subset of E_n and c is a real number, y a vector, $cS + y$ denotes the set

$$\{cx + y : x \in S\}$$

10. The fundamental theorem of Minkowski

We now consider the lattice G, consisting of the points g of E_n with integral coordinates g_k. Let K be a convex body with center 0. Then K has a volume V in the Peano–Jordan sense, as was already shown by

Minkowski. The theorem of Minkowski that follows is the fundamental theorem of the geometry of numbers.

Theorem VI.4. If $V > 2^n$ then K contains a point with integral coordinates, other than 0.

Proof. Suppose that K contains no point with integral coordinates, other than 0. Contract K towards the origin in the linear ratio 2: 1 forming the new convex body $\frac{1}{2}K$ with center 0. Clearly,

$$V(\tfrac{1}{2}K) = V(K)/2^n = V/2^n$$

Place around each point $g \in G$ the body $\frac{1}{2}K_g$, derived from $\frac{1}{2}K$ by the translation which moves 0 to g. Thus

$$\tfrac{1}{2}K_g = \tfrac{1}{2}K + g$$

These bodies do not overlap. To prove this, suppose that g, g' are distinct points of G such that $\frac{1}{2}K_g$ and $\frac{1}{2}K_{g'}$ have a point p in common. Then $p \in \frac{1}{2}K + g, p \in \frac{1}{2}K + g'$. Thus there are points x, y of K such that

$$p = \tfrac{1}{2}x + g = \tfrac{1}{2}y + g'$$

which implies that

$$(x - y)/2 = g' - g.$$

Since K is convex and has center 0, $(x - y)/2$, and hence $g' - g$, belongs to K. But $g' - g$ is a nonzero point of G. This contradicts the assumption that K contains no points of G other than 0. Hence the bodies do not overlap.

Consider now the bodies $\frac{1}{2}K_g$ whose centers satisfy

$$0 \leq g_k \leq n - 1, \qquad 1 \leq k \leq n$$

There are N^n such bodies and their total volume is

$$N^n V(\tfrac{1}{2}K) = N^n V/2^n$$

We can find a number c, depending on $\frac{1}{2}K$ but *not on* N, such that all these bodies are contained in a cube of side $N + 2c$. Hence

$$N^n\, V/2^n \leq (N + 2c)^n,$$

$$V/2^n \leq (1 + 2c/N)^n$$

Taking the limit as $N \longrightarrow \infty$, we obtain

$$V/2^n \leq 1, \qquad V \leq 2^n$$

This contradicts the hypothesis of the theorem and completes the proof.

11. Another form of the fundamental theorem

A restatement of the fundamental theorem is

Theorem VI.5. If K is convex and has center 0, with volume $V \geq 2^n$, then K contains a point with integral coordinates other than 0 either in its interior or on its boundary.

Proof. We apply Theorem VI.4 to the body $(1 + \varepsilon)K$, $\varepsilon > 0$ and let $\varepsilon \longrightarrow 0$.

12. Proof that an ellipsoid is convex and centrally symmetric

Our first application of the fundamental theorem will be to the determination of a bound for the arithmetic minimum of a positive quadratic form. Let A be a real symmetric positive definite matrix,

$$d = \det(A),$$

and let E be the ellipsoid

$$x^{\mathrm{T}}Ax \leq c \tag{11}$$

Let U be a real matrix such that $\det(U) = 1$ and

$$A = U^{\mathrm{T}}DU, \qquad D = \operatorname{diag}(d_1, d_2, \ldots, d_n), \qquad d_k > 0$$

Then the transformation

$$y = Ux \tag{12}$$

takes the ellipsoid E into the ellipsoid E' defined by $y^{\mathrm{T}}Dy \leq c$. Put

$$z_k = (d_k/c)^{1/2}y_k, \qquad 1 \leq k \leq n \tag{13}$$

Then the ellipsoid E' becomes the unit sphere S_n. Thus to show that E is convex and has center 0 it is only necessary to show that S_n is convex and has center 0.

13. Proof that the unit sphere is convex

That S_n has center 0 is obvious. To prove that S_n is convex, let x_k, y_k satisfy

$$\sum_{k=1}^{n} x_k^2 \leq 1, \qquad \sum_{k=1}^{n} y_k^2 \leq 1$$

Suppose that $a, b \geq 0$, $a + b = 1$. Then

$$\begin{aligned}\sum_{k=1}^{n} (ax_k + by_k)^2 &= a^2 \sum_{k=1}^{n} x_k^2 + 2ab \sum_{k=1}^{n} x_k y_k + b^2 \sum_{k=1}^{n} y_k^2 \\ &\leq a^2 \sum_{k=1}^{n} x_k^2 + ab \sum_{k=1}^{n} (x_k^2 + y_k^2) + b^2 \sum_{k=1}^{n} y_k^2 \\ &\leq a^2 + 2ab + b^2 = 1\end{aligned}$$

Thus S_n is convex; and we have proved

Lemma VI.1. The ellipsoid (11) is convex and has center 0.

14. The volume of an ellipsoid

By means of the transformations (12) and (13) we can compute $V(E)$. The Jacobian of the transformation (12) is 1, and the Jacobian of the transformation (13) is

$$(d_1 d_2 \cdots d_n)^{1/2}/c^{n/2} = d^{1/2}/c^{n/2}$$

Thus

$$V(E) = \frac{c^{n/2}}{d^{1/2}} V(S_n)$$

and it is only necessary to compute $V(S_n)$. We have

$$V(S_n) = \int_{x_1^2 + x_2^2 + \cdots + x_n^2 \leq 1} dx_1\, dx_2 \cdots dx_n$$

Put

$$x_k = (1 - x_n^2)^{1/2} y_k, \qquad 1 \leq k \leq n-1, \qquad x_n = y_n$$

Then

$$\begin{aligned}V(S_n) &= \int_{-1}^{1} (1 - y_n^2)^{(n-2)/2}\, dy_n \int_{y_1^2 + y_2^2 + \cdots + y_{n-1}^2 \leq 1} dy_1\, dy_2 \cdots dy_{n-1} \\ &= \frac{\Gamma(\frac{1}{2}(n+1))\Gamma(\frac{1}{2})}{\Gamma(\frac{1}{2}(n+2))} V(S_{n-1})\end{aligned}$$

It follows that

(14) $$V(S_n) = \pi^{n/2}/\Gamma(1 + n/2)$$

Formula (14) now implies

Lemma VI.2. The volume of the ellipsoid E defined by (11) is just

(15) $$V(E) = (\pi c)^{n/2}/d^{1/2}\Gamma(1 + n/2)$$

15. An application to the arithmetic minimum

We are now ready to apply Minkowski's fundamental theorem (in the form of Theorem VI.5) to E. Let c be chosen so that $V(E) \geq 2^n$, giving

$$c \geq \frac{4}{\pi}\Gamma\left(1 + \frac{n}{2}\right)^{2/n} d^{1/n}$$

Then E must have a point with integral coordinates, other than 0, either in its interior or on its boundary. Hence we have proved

Theorem VI.6. The arithmetic minimum of the positive definite form $x^{\mathrm{T}}Ax$ does not exceed

(16) $$\frac{4}{\pi}\Gamma\left(1 + \frac{n}{2}\right)^{2/n} d^{1/n}$$

We note that Stirling's formula shows that

$$\frac{4}{\pi}\Gamma\left(1 + \frac{n}{2}\right)^{2/n} \sim \frac{2n}{\pi e}$$

The bound given by Hermite was $(\frac{4}{3})^{(n-1)/2}$ which grows exponentially. Minkowski's result therefore is much better for large n.

16. Restatement of the fundamental theorem

Let K be convex with center 0 and let P map K onto K', where the columns of P are a lattice basis of L, so that $|\det(P)| = \Delta$. Let $V = V(K)$, $V' = V(K')$. Then $V' = \Delta V$. The requirement

$$V > 2^n$$

is completely equivalent to the requirement

$$V' > 2^n\Delta \tag{17}$$

If (17) is satisfied, then K must contain a point other than 0 with integral coordinates. Thus K' must contain the image of this point under P, which is a point of L other than 0. Thus another version of Minkowski's result is

Theorem VI.7. A convex body with center 0 and volume $> 2^n\Delta$ contains a point other than 0 of each lattice L of determinant Δ.

17. Minkowski's "linear forms" theorem

Some applications to particular bodies will now be considered. Let B be the "box" defined by

$$|x_k| < \lambda_k, \qquad 1 \le k \le n$$

Then B is convex with center 0, and $V(B) = 2^n\lambda_1\lambda_2 \cdots \lambda_n$. Thus a lattice L of determinant Δ must necessarily have a point $\neq 0$ in B, provided that

$$V(B) > 2^n\Delta$$

or that

$$\lambda_1\lambda_2 \cdots \lambda_n > \Delta$$

This result has the following formulation.

Theorem VI.8. Let

$$\xi_i = \sum_{k=1}^{n} p_{ik}g_k, \qquad 1 \le i \le n \tag{18}$$

be real linear forms in $g_1, g_2, \ldots, g_n$, $\Delta = |\det(p_{ij})| \neq 0$. Let $\lambda_1, \lambda_2, \ldots, \lambda_n$ be positive numbers with

$$\lambda_1\lambda_2 \cdots \lambda_n > \Delta$$

Then there exist integers $g_1, g_2, \ldots, g_n$ not all 0 such that

$$|\xi_i| < \lambda_i, \qquad 1 \le i \le n$$

This is the famous "linear forms theorem" of Minkowski.

18. Generalization to complex forms

A generalization to complex linear forms is possible, provided that these occur in conjugate complex pairs. Suppose that the set $\{\xi_i\}$ contains r real forms and s pairs of conjugate complex forms, so that $r + 2s = n$. Let Δ (as before) be the absolute value of the determinant of the forms. Let $\lambda_1, \lambda_2, \ldots, \lambda_n$ be positive numbers such that $\lambda_\alpha = \lambda_\beta$ whenever ξ_α and ξ_β are conjugate complex forms, and $\lambda_1\lambda_2 \cdots \lambda_n > \Delta$.

We construct a new set of n real forms ξ_i' and a new set of n positive numbers λ_i' as follows: If ξ_i is a real form, define $\xi_i' = \xi_i$ and $\lambda_i' = \lambda_i$; but if ξ_α, ξ_β are a pair of conjugate complex forms, define

$$\xi_\alpha' = \frac{\xi_\alpha + \xi_\beta}{2}, \qquad \xi_\beta' = \frac{\xi_\alpha - \xi_\beta}{2i}$$

$$\lambda_\alpha' = \lambda_\beta' = \lambda_\alpha/\sqrt{2}$$

Then Δ' (the absolute value of the determinant of the forms ξ_i') is just $2^{-s}\Delta$, and

$$\lambda_1'\lambda_2' \cdots \lambda_n' = 2^{-s}\lambda_1\lambda_2 \cdots \lambda_n > 2^{-s}\Delta = \Delta'$$

Applying Theorem VI.8 to the forms ξ_i' and the numbers λ_i', we find that integers $g_1, g_2, \ldots, g_n$ exist such that

$$|\xi_k'| < \lambda_k', \qquad 1 \le k \le n \tag{19}$$

If ξ_k is a real form then (19) becomes

$$|\xi_k| < \lambda_k$$

If ξ_α, ξ_β are a pair of conjugate complex forms then (19) implies that

$$|\xi_\alpha|^2 = |\xi_\alpha'|^2 + |\xi_\beta'|^2 < \frac{\lambda_\alpha^2}{2} + \frac{\lambda_\alpha^2}{2} = \lambda_\alpha^2, \qquad |\xi_\alpha| < \lambda_\alpha$$

Hence Theorem VI.8 may be generalized in the direction indicated.

19. An application to simultaneous diophantine approximation

An interesting application of Theorem VI.8 arises from the choice

$$\xi_k = g_k - r_k g_n, \qquad 1 \le k \le n-1, \qquad \xi_n = g_n/t^n$$

where the r_k are arbitrary real numbers and $t > 0$. Then

$$P = \begin{bmatrix} 1 & 0 & 0 & \cdots & 0 & -r_1 \\ 0 & 1 & 0 & \cdots & 0 & -r_2 \\ 0 & 0 & 1 & \cdots & 0 & -r_3 \\ \cdot & \cdot & \cdot & & \cdot & \cdot \\ \cdot & \cdot & \cdot & & \cdot & \cdot \\ \cdot & \cdot & \cdot & & \cdot & \cdot \\ 0 & 0 & 0 & \cdots & 1 & -r_{n-1} \\ 0 & 0 & 0 & \cdots & 0 & 1/t^n \end{bmatrix}$$

so that $\det(P) = 1/t^n$, $\Delta = 1/t^n$. Now choose $\lambda_k = \lambda$, $1 \leq k \leq n$, so that $\lambda_1 \lambda_2 \cdots \lambda_n > \Delta$; that is, $\lambda > 1/t$. Then Theorem VI.8 assures us that integers $g_1, g_2, \ldots, g_n$ not all zero exist such that

$$|r_k g_n - g_k| < \lambda, \qquad 1 \leq k \leq n-1, \qquad |g_n| < \lambda t^n$$

This is one of Dirichlet's theorems on simultaneous diophantine approximation.

The limiting case $\lambda = 1/t$ is of independent interest and yields the following (here n is replaced by $n + 1$):

Theorem VI.9. Let $r_1, r_2, \ldots, r_n$ be arbitrary real numbers. Let t be any number > 1. Then integers $p_1, p_2, \ldots, p_n, q$ exist such that $1 \leq q \leq t^n$ and

(20) $$r_k = \frac{p_k}{q} + \frac{\theta_k}{qt}, \qquad |\theta_k| \leq 1, \quad 1 \leq k \leq n$$

20. Proof of the theorem on lattices

We are now in a position to supply a proof of Theorem VI.1, with the aid of Theorem VI.9 above. Certainly any lattice L satisfies requirements (2), (3), (4). Suppose conversely that L is a subset of E_n satisfying these requirements. By (4), we may assume that a positive number ε exists such that the sphere $\|\xi\| < \varepsilon$ meets L in precisely 0. Let $\xi_1, \xi_2, \ldots, \xi_n$ be n linearly independent vectors of L, whose existence is guaranteed by (3). Choose $t > 0$ so large that

(21) $$(1/t) \sum_{k=1}^{n} \|\xi_k\| < \varepsilon$$

Put

$$L_0 = \{\xi_1, \xi_2, \ldots, \xi_n\}$$

Let ξ be any element of L. Then certainly real numbers $r_1, r_2, \ldots, r_n$ exist such that

$$\xi = \sum_{k=1}^{n} r_k \xi_k$$

since the ξ_k span E_n. By Theorem VI.9 above we can determine integers $p_1, p_2, \ldots, p_n, q$ such that (20) holds. Then

$$q\xi = \sum_{k=1}^{n} p_k \xi_k + (1/t) \sum_{k=1}^{n} \theta_k \xi_k = \eta + \zeta$$

where $\eta \in L_0$. By (2),

$$\zeta = q\xi - \eta \in L$$

and by (21),

$$\|\zeta\| < \varepsilon$$

It follows that $\zeta = 0$. Thus $q\zeta \in L_0$. Since the number of choices for q is bounded ($1 \leq q \leq t^n$) if follows that a fixed positive integer m exists such that

$$m\xi \in L_0, \qquad \text{all} \quad \xi \in L$$

Hence L is contained in the lattice

$$\frac{1}{m} L_0 = \left\{ \frac{\xi_1}{m}, \frac{\xi_2}{m}, \ldots, \frac{\xi_n}{m} \right\}$$

and consequently must itself be a lattice. This completes the proof.

21. The octahedron

For our next application we consider the "octahedron" C defined by

$$|x_1| + |x_2| + \cdots + |x_n| \leq \lambda$$

Clearly C has center 0 and is convex. We have

$$V(C) = \int_{|x_1|+|x_2|+\cdots+|x_n| \leq \lambda} dx_1\, dx_2 \cdots dx_n$$

$$= 2^n \lambda^n \int_{x_1+x_2+\cdots+x_n \leq 1,\, x_k \geq 0} dx_1\, dx_2 \cdots dx_n$$

Set $V(C) = 2^n \lambda^n c_n$. Then the substitution

$$x_k = (1 - x_n) y_k, \qquad 1 \leq k \leq n-1, \quad x_n = y_n$$

shows that $c_n = (1/n)c_{n-1}$, so that $c_n = 1/n!$, since $c_1 = 1$. Thus

$$V(C) = 2^n\lambda^n/n!$$

Therefore if λ is chosen so that

$$V(C) \geq 2^n\Delta, \qquad \lambda \geq (n!\,\Delta)^{1/n}$$

then any lattice L of determinant Δ must have a point other than 0, in or on C. This gives another theorem on linear forms.

22. Further results on linear forms

Theorem VI.10. Let

$$\xi_i = \sum_{k=1}^{n} p_{ik}g_k, \qquad 1 \leq i \leq n$$

be linear forms in $g_1, g_2, \ldots, g_n$, $\Delta = |\det(p_{ij})| \neq 0$. Then there are integers $g_1, g_2, \ldots, g_n$ not all zero such that

$$|\xi_1| + |\xi_2| + \cdots + |\xi_n| \leq (n!\,\Delta)^{1/n}$$

This theorem has the following corollary:

Corollary VI.1. Let

$$\xi_i = \sum_{k=1}^{n} p_{ik}g_k, \qquad 1 \leq i \leq n$$

be linear forms in $g_1, g_2, \ldots, g_n$, $\Delta = |\det(p_{ij})| \neq 0$. Then there are integers $g_1, g_2, \ldots, g_n$ not all zero such that

(22) $$|\xi_1\xi_2 \cdots \xi_n| \leq (n!/n^n)\Delta$$

The proof is a simple application of the arithmetic–geometric mean inequality.

23. Consequences of interest in algebraic number theory

Corollary VI.1 also can be generalized to the case when the forms ξ_i are complex, provided that these occur in conjugate complex pairs. Theorem VI.10 is applied to the forms ξ_i', the arithmetic–geometric mean inequality is applied once again, and the result (22) is obtained, except that the right side becomes

$$(2^s n!/n^n)\Delta$$

This result is fundamental in algebraic number theory. If d is the discriminant of an algebraic number field K of degree n then a slight refinement of the previous argument shows that

$$|d| \geq \left\{\left(\frac{\pi}{4}\right)^s \frac{n^n}{n!}\right\}^2 \tag{23}$$

where $2s$ is the number of conjugate complex fields associated with K. (23) shows in particular that $|d| \longrightarrow \infty$ as $n \longrightarrow \infty$.

24. A general figure

We consider now the more general figure C_σ defined by

$$\{|x_1|^\sigma + |x_2|^\sigma + \cdots + |x_n|^\sigma\}^{1/\sigma} \leq \lambda, \qquad \sigma \geq 1$$

$\sigma = 1$ gives the octahedron and $\sigma = 2$ the sphere; while $\sigma \longrightarrow \infty$ gives the cube. To prove that C_σ is convex it is necessary to know that

$$\frac{a+b}{2} \leq \left\{\frac{a^\sigma + b^\sigma}{2}\right\}^{1/\sigma} \qquad \text{for} \quad \sigma \geq 1$$

It is trivial that C_σ has center 0. To compute $V(C_\sigma)$, we have that $V(C_\sigma) = 2^n\lambda^n v_n$, where

$$v_n = \int dx_1\, dx_2 \cdots dx_n, \qquad x_1^\sigma + x_2^\sigma + \cdots + x_n^\sigma \leq 1, \quad x_k \geq 0$$

As before, the substitution

$$x_k = (1 - x_n^\sigma)^{1/\sigma} y_k, \qquad 1 \leq k \leq n-1, \quad x_n = y_n$$

produces

$$v_n = \frac{\Gamma(1 + (n-1)/\sigma)}{\Gamma(1 + n/\sigma)} \Gamma\left(1 + \frac{1}{\sigma}\right) v_{n-1}$$

from which we find that

$$v_n = \frac{\Gamma(1 + 1/\sigma)^n}{\Gamma(1 + n/\sigma)},$$

$$V(C_\sigma) = 2^n\lambda^n \frac{\Gamma(1 + 1/\sigma)^n}{\Gamma(1 + n/\sigma)} \tag{24}$$

Now choose λ so that

$$V(C_\sigma) \geq 2^n\Delta,$$

$$\lambda \geq \frac{\Gamma(1 + n/\sigma)^{1/n}}{\Gamma(1 + 1/\sigma)} \Delta^{1/n}$$

Then any lattice L of determinant Δ must have a point other than 0 in C_σ. Thus we have

Theorem VI.11. Let

$$\xi_i = \sum_{k=1}^{n} p_{ik} g_k, \qquad 1 \leq i \leq n$$

be linear forms in $g_1, g_2, \ldots, g_n$, $\Delta = |\det(p_{ij}| \neq 0$. Then there are integers $g_1, g_2, \ldots, g_n$ not all zero such that

$$\{|\xi_1|^\sigma + |\xi_2|^\sigma + \cdots + |\xi_n|^\sigma\}^{1/\sigma} \leq \frac{\Gamma(1 + n/\sigma)^{1/n}}{\Gamma(1 + 1/\sigma)} \Delta^{1/n}$$

Chapter VII

Groups of Matrices

It goes without saying that the selection of material for this chapter must be highly limited and individualistic, since the subject is so vast as to encompass practically the whole of mathematics.

1. Definition of $GL(n, S)$, $SL(n, S)$

We introduce new notation at this point, to conform with current usage. Let S be any commutative ring with a unit element 1. Then the group of unit matrices S_n will be denoted by $GL(n, S)$ (GL standing for "general linear group") instead of S_n'. The subgroup of $GL(n, S)$ consisting of all matrices of determinant 1 will be denoted by $SL(n, S)$ (SL standing for "special linear group").

2. Some properties of $GL(n, S)$, $SL(n, S)$

It is clear that $SL(n, S)$ is a normal subgroup of $GL(n, S)$, and that a complete set of coset representatives for $GL(n, S)$ modulo $SL(n, S)$ consists of the matrices

$$\varepsilon I_1 \dotplus I_{n-1}$$

where ε runs over all units of S. Furthermore, it is clear that $SL(n, S) \supset GL'(n, S)$, the commutator subgroup of $GL(n, S)$, and hence $GL(n, S)/SL(n, S)$ is abelian. Thus $SL(n, S)$ is of finite index in $GL(n, S)$ if and only if S contains finitely many units, and the total number of units of S, if finite, is equal to this index. For example if $S = Z$ then the index is 2; and if $S = GF(q)$, the finite field with q elements, the index is $q - 1$.

3. The centers of $GL(n, S)$, $SL(n, S)$

We first determine the centers of $GL(n, S)$, $SL(n, S)$, which we denote by $CGL(n, S)$, $CSL(n, S)$, respectively.

Theorem VII.1. Let S be a commutative ring with a unit element 1. Then the center of $GL(n, S)$ consists of all scalar matrices εI, where ε is a unit of S.

Proof. We may assume that $n > 1$. Let p, q be integers such that $1 \leq p, q \leq n, p \neq q$. Then the matrices

$$I + E_{pq}, \tag{1}$$

$$I - E_{pp} + E_{pq} - E_{qp} \tag{2}$$

each belong to $SL(n, S)$.

Suppose that $A = (a_{ij})$ is any element of $CGL(n, S)$. Then A commutes with each of the matrices (1) and (2), which implies that A commutes with E_{pq} and E_{pp}. Thus A commutes with E_{pq} for *all* integers p, q such that $1 \leq p, q \leq n$. Since $E_{pq} = (\delta_{ip}\delta_{jq})$, a brief calculation shows that

$$\delta_{ip}a_{qj} = \delta_{jq}a_{ip}$$

Choose $i = p$. Then

$$a_{qj} = \delta_{jq}a_{pp}$$

so that $a_{qj} = 0$ if $j \neq q$, and $a_{qq} = a_{pp}$ if $j = q$. Hence A is scalar. Put $A = \varepsilon I$, where ε belongs to S. Then $\det(A) = \varepsilon^n$ is a unit of S, so that ε must also be a unit of S. Conversely, any such matrix clearly belongs to $CGL(n, S)$. This completes the proof.

Since the matrices (1) and (2) belong to $SL(n, S)$, this proof also allows us to determine the center of $SL(n, S)$:

Theorem VII.2 Let S be a commutative ring with a unit element 1. Then the center of $SL(n, S)$ consists of all scalar matrices εI, where ε is a unit of S such that $\varepsilon^n = 1$.

These theorems show that there is a one-to-one correspondence between the units of S and the center of $GL(n, S)$. When S is a domain of integrity, Theorem VII.2 shows that the center of $SL(n, S)$ must be finite and of order at most n, since in a domain of integrity a polynomial equation of degree p can have at most p solutions. For example if $S = Z$ then the center of $SL(n, S)$ is of order $(n, 2)$; and if $S = GF(q)$, the center of $SL(n, S)$ is of order $(n, q - 1)$, since every nonzero element of $GF(q)$ is of order $q - 1$.

4. Definition of $PGL(n, S)$, $PSL(n, S)$

We define

$$PGL(n, S) = GL(n, S)/CGL(n, S),$$
$$PSL(n, S) = SL(n, S)/CSL(n, S)$$

so that these quotient groups have trivial center. The P stands for "projective." The notation $LF(n, S)$ for $PSL(n, S)$ is also frequently encountered (LF stands for "linear fractional").

5. The generators of $SL(n, R)$, R a euclidean ring which is not of characteristic 2

We now consider the generators of $SL(n, R)$ when R is a euclidean ring which is not of characteristic 2.

If we refer to the proof of the theorem of the Hermite normal form given in Chapter II, we see that any element A of $GL(n, R)$ may be written as $A = BC$, where B is a diagonal matrix whose nonzero entries are units of R, and C belongs to the group generated by the permutation matrices and the matrices

$$S(\mu) = \begin{bmatrix} 1 & \mu \\ 0 & 1 \end{bmatrix} \dotplus I_{n-2}, \qquad \mu \in R$$

If in addition A belongs to $SL(n, R)$, then $\det(B) = \pm 1$, since $\det(A) = 1$, $\det(C) = \pm 1$. An easy calculation shows that B belongs to the group generated by the permutation matrices, the matrix

$$E = -I_1 \dotplus I_{n-1}$$

and the matrices

$$D(\varepsilon) = \begin{bmatrix} \varepsilon & 0 \\ 0 & 1/\varepsilon \end{bmatrix} \dotplus I_{n-2}, \qquad \varepsilon \in R'$$

It follows that $SL(n, R)$ is generated by the following matrices (not all of which belong to $SL(n, R)$):

$$U = \begin{bmatrix} 0 & 1 \\ 1 & 0 \end{bmatrix} \dotplus I_{n-2}, \qquad V = \begin{bmatrix} 0 & 1 & 0 & \cdots & 0 \\ 0 & 0 & 1 & \cdots & 0 \\ \cdot & \cdot & \cdot & & \cdot \\ \cdot & \cdot & \cdot & & \cdot \\ \cdot & \cdot & \cdot & & \cdot \\ 1 & 0 & 0 & \cdots & 1 \\ 0 & 0 & 0 & \cdots & 0 \end{bmatrix},$$

$$E, \qquad S(\mu), \quad \mu \in R, \qquad D(\varepsilon), \quad \varepsilon \in R'$$

The matrices U, V are well-known generators of the group of permutation matrices.

This list of generators may be reduced. First we note the identity

$$D(\varepsilon) = UES\left(-\frac{1}{\varepsilon}\right)US(\varepsilon)US\left(-\frac{1}{\varepsilon}\right), \qquad \varepsilon \in R',$$

so that these may be dropped from the list. Next, in order to obtain a set of generators which belong to $SL(n, R)$, we replace U, V by T, P, respectively, where

$$T = UE = \begin{bmatrix} 0 & 1 \\ -1 & 0 \end{bmatrix} \dotplus I_{n-2},$$

$$P = \begin{cases} U, & n \text{ odd} \\ UE, & n \text{ even} \end{cases}$$

If we now note the identitites

$$ETE^{-1} = T^{-1},$$
$$EPE^{-1} = PT^2,$$
$$ES(\mu)E^{-1} = S(-\mu), \qquad \mu \in R$$

we see that any element A of $SL(n, R)$ may be written as B or EB, where

B belongs to the group generated by T, P, and the matrices $S(\mu)$, $\mu \in R$. But both A and B have determinant 1, and so A must equal B. Finally, L. K. Hua and I. Reiner have shown [18] that T belongs to the group generated by P and $S(1)$, by means of an elementary but complicated identity.

Summarizing, we have proved

Theorem VII.3. If R is a euclidean ring which is not of characteristic 2, then $SL(n, R)$ is generated by

$$P = \begin{bmatrix} 0 & 1 & 0 & \cdots & 0 \\ 0 & 0 & 1 & \cdots & 0 \\ \cdot & \cdot & \cdot & & \cdot \\ \cdot & \cdot & \cdot & & \cdot \\ \cdot & \cdot & \cdot & & \cdot \\ 0 & 0 & 0 & & 1 \\ (-1)^{n-1} & 0 & 0 & \cdots & 0 \end{bmatrix}$$

$$S(\mu) = \begin{bmatrix} 1 & \mu \\ 0 & 1 \end{bmatrix} \dotplus I_{n-2}, \qquad \mu \in R$$

These generators may be reduced further when R has additional properties. Thus if R is a finitely generated Z-module with generators $\mu_1, \mu_1, \ldots, \mu_r$, then $SL(n, R)$ is also finitely generated with generators P, $S(\mu_i)$, $1 \leq i \leq r$. For example if $R = Z$ then $SL(n, R)$ is generated by P, $S(1)$; and if $R = GF(q)$, where $q = p^k$, p prime, then $SL(n, R)$ is generated by P, $S(\theta^i)$, $0 \leq i \leq k - 1$, where θ is any element of $GF(q)$ such that $GF(q) = GF(p)(\theta)$.

6. The commutator subgroup $GL'(n, R)$

Because of Theorem VII.3, the commutator subgroup $GL'(n, R)$ of $GL(n, R)$ is easily determined. We first note the identity

$$\begin{bmatrix} 1 & \mu & 0 \\ 0 & 1 & 0 \\ 0 & 0 & 1 \end{bmatrix} = \begin{bmatrix} 1 & 0 & \mu \\ 0 & 1 & 0 \\ 0 & 0 & 1 \end{bmatrix} \begin{bmatrix} 1 & 0 & 0 \\ 0 & 1 & 0 \\ 0 & 1 & 1 \end{bmatrix} \begin{bmatrix} 1 & 0 & \mu \\ 0 & 1 & 0 \\ 0 & 0 & 1 \end{bmatrix}^{-1} \begin{bmatrix} 1 & 0 & 0 \\ 0 & 1 & 0 \\ 0 & 1 & 1 \end{bmatrix}^{-1}$$

so that if $n > 2$, $S(\mu)$ belongs to $SL'(n, R)$ for all $\mu \in R$. As for P, it was also shown by Hua and Reiner [19] that P belongs to the commutator subgroup of the group of generalized permutation matrices. From this it follows that $SL(n, R) \subset GL'(n, R)$; and since $GL'(n, R)$ is trivially contained in $SL(n, R)$, we get

Theorem VII.4. If R is a euclidean ring which is not of characteristic 2, and $n > 2$, then $SL(n, R) = GL'(n, R)$.

7. The commutator subgroup $SL'(n, R)$

As the proof of this theorem suggests, the following result is also true:

Theorem VII.5. If R is a euclidean ring which is not of characteristic 2, and $n > 2$, then $SL'(n, R) = SL(n, R)$.

Proof. It is only necessary to show that the matrix P belongs to $SL'(n, R)$. Suppose first that n is odd. Then P is a permutation matrix and, as a permutation, is an n-cycle. Since n is odd P^2 is also an n-cycle, and so a permutation matrix Q exists such that $P^2 = Q^{-1}PQ$, whence $P = P^{-1}Q^{-1}PQ$. If $\det(Q) = 1$, this shows that P belongs to $SL'(n, R)$. If $\det(Q) = -1$, then $\det(-Q) = 1$ (since n is odd) and the identity $P = P^{-1}(-Q)^{-1}P(-Q)$ shows that P belongs to $SL'(n, R)$.

Now suppose that n is even. Write $P = P_n$. Then we verify easily that

$$P_n T^{-1} = I_1 \dotplus P_{n-1}$$

Since $n - 1$ is odd, the previous discussion implies that $P_n T^{-1}$ belongs to $SL'(n, R)$. It is thus sufficient to show that T is in $SL'(n, R)$. But we have the identity

$$\begin{bmatrix} 0 & 1 & 0 \\ -1 & 0 & 0 \\ 0 & 0 & 1 \end{bmatrix} = \begin{bmatrix} 1 & 1 & 0 \\ 0 & 1 & 0 \\ 0 & 0 & 1 \end{bmatrix} \begin{bmatrix} 1 & 0 & 0 \\ -1 & 1 & 0 \\ 0 & 0 & 1 \end{bmatrix} \begin{bmatrix} 1 & 1 & 0 \\ 0 & 1 & 0 \\ 0 & 0 & 1 \end{bmatrix}$$

By the argument of Theorem VII.4

$$\begin{bmatrix} 1 & 1 & 0 \\ 0 & 1 & 0 \\ 0 & 0 & 1 \end{bmatrix}$$

belongs to $SL'(3, R)$. Similarly,

$$\begin{bmatrix} 1 & 0 & 0 \\ -1 & 1 & 0 \\ 0 & 0 & 1 \end{bmatrix}$$

also belongs to $SL'(3, R)$. Thus if $n > 2$, T belongs to $SL'(n, R)$. This completes the proof.

The case $n = 2$ is exceptional.

We now return to the principal ideal ring R. A particularly important class of subgroups of $GL(n, R)$ are the so-called *congruence* groups, which may be defined in various ways. For purposes of simplicity we shall limit ourselves to $SL(n, R)$; and in order to avoid cumbersome notation, we write

$$\Gamma = SL(n, R)$$

8. Principal congruence groups

Let μ be any nonzero element of R. Then the *principal congruence subgroup* of Γ of *level* μ is the set of all matrices A of Γ such that

$$A \equiv I \bmod \mu$$

where congruence is elementwise congruence. We denote this subgroup by $\Gamma(\mu)$.

9. Quotient groups and an isomorphism

Clearly, $\Gamma(\mu)$ is a normal subgroup of Γ. Quite often it is the quotient group

$$G(\mu) = \Gamma/\Gamma(\mu)$$

which is of interest, and in this connection the following result is valuable:

Theorem VII.6. We have the isomorphism

$$G(\mu) = \Gamma/\Gamma(\mu) \cong SL(n, R/(\mu))$$

Proof. Let φ be the natural homomorphism of R determined by (μ), so that φ maps every element of R into its representative in some complete set of residues modulo μ. Thus

$$\varphi: R \longrightarrow R/(\mu)$$

Then φ induces a homomorphism of $SL(n, R)$ onto $SL(n, R/(\mu))$. Let K be the kernel of this homomorphism. Clearly, K must consist of all matrices A of $SL(n, R)$ such that $A \equiv I \bmod \mu$, so that $K = \Gamma(\mu)$. It follows that

$$SL(n, R)/\Gamma(\mu) \cong SL(n, R/(\mu))$$

This completes the proof.

We see then that $\Gamma(\mu)$ is of finite index in Γ if and only if $R/(\mu)$ is finite [equivalently, if and only if $N(\mu)$ is finite].

10. Congruence groups

We now define a *congruence group* as any subgroup of Γ which contains a principal congruence group. The importance of the congruence groups stems from the fact that in many important instances, *any* subgroup of Γ of finite index is a congruence group. This has been proved by Mennicke [32] for the case when $R = Z$ and $n \geq 3$ (it is false for $n = 2$). Another proof is given by Bass, Lazard, and Serre [2], and far-reaching generalizations have been given by Bass, Milnor, and Serre.

11. Structure theorems for $\Gamma(\mu)$

We first prove some lemmas and theorems whose object is to reduce the study of the quotient groups $G(\mu)$ to the case when μ is a prime power. The results that follow were proved for the case $R = Z$ [54]. We let μ, ν be arbitrary nonzero elements of R, and we put

$$(\mu, \nu) = \delta, \qquad [\mu, \nu] = \mu\nu/\delta = \Delta$$

so that $[\mu, \nu]$ is the least common multiple of μ and ν.

Lemma VII.1. Suppose that A belongs to $\Gamma(\delta)$. Then an element X of Γ can be determined so that

(3) $$X \equiv I \bmod \mu,$$

(4) $$X \equiv A \bmod \nu$$

Proof. Since $A \in \Gamma(\delta)$ we can write $A = I + \delta B$, where $B \in R_n$. Set $X = I + \mu Y$. Then (3) is satisfied, and (4) becomes

$$\mu Y \equiv \delta B \bmod \nu, \qquad (\mu/\delta) Y \equiv B \bmod \nu/\delta$$

Since $(\mu/\delta, \nu/\delta) = 1$, this has a solution Y. Hence there is an X_0 satisfying (3) and (4). Now determine X so that $X \equiv X_0 \bmod \Delta$, $\det(X) = 1$, which we may do by (b) of Section II.21. Then this X satisfies the conditions of the lemma.

This lemma implies

Lemma VII.2. Let A belong to $\Gamma(\delta)$. Then B and C may be found such that $B \in \Gamma(\mu)$, $C \in \Gamma(\nu)$, and $A = BC$.

Proof. Determine X as in Lemma VII.1. Choose $B = X$, $C = X^{-1}A$.

These lemmas yield the following structure theorem:

Theorem VII.7. $\Gamma(\mu)$, $\Gamma(\nu)$ are normal subgroups of $\Gamma(\delta)$, and

(5) $$\Gamma(\mu)\Gamma(\nu) = \Gamma(\delta),$$

(6) $$\Gamma(\mu) \cap \Gamma(\nu) = \Gamma(\Delta)$$

Proof. It is clear that $\Gamma(\mu)$, $\Gamma(\nu)$ are normal subgroups of $\Gamma(\delta)$, and that $\Gamma(\mu)\Gamma(\nu) \subset \Gamma(\delta)$. Lemma VII.2 implies that $\Gamma(\mu)\Gamma(\nu) \supset \Gamma(\delta)$. Thus (5) is proved. Furthermore both $\Gamma(\mu)$ and $\Gamma(\nu)$ contain $\Gamma(\Delta)$, so that $\Gamma(\mu) \cap \Gamma(\nu) \supset \Gamma(\Delta)$. Suppose that $A \in \Gamma(\mu) \cap \Gamma(\nu)$. Then $A \equiv I \bmod \mu$, $A \equiv I \bmod \nu$, which implies that $A \equiv I \bmod \Delta$. Hence $\Gamma(\mu) \cap \Gamma(\nu) \subset \Gamma(\Delta)$ and so (6) is proved.

We now get

Theorem VII.8. We have the isomorphism

(7) $$\Gamma(\delta)/\Gamma(\mu) \cong \Gamma(\nu)/\Gamma(\Delta)$$

Proof. Set $G = \Gamma(\mu)$, $H = \Gamma(\nu)$. By one of the isomorphism theorems

$$GH/G \cong H/G \cap H$$

The result now follows from Theorem VII.7.

12. Isomorphism theorems for $G(\alpha, \beta)$

In order to get at the structure of the quotient groups $G(\mu)$, we introduce the groups

$$G(\alpha, \beta) = \Gamma(\alpha)/\Gamma(\beta)$$

where α, β are elements of R such that $\alpha \mid \beta$. Then (7) becomes

$$G(\delta, \mu) \cong G(\nu, \Delta) \tag{8}$$

From Theorem VII.8 we conclude

Theorem VII.9. Let $\times$ represent direct product. Then

$$G(\delta, \Delta) \cong G(\delta, \mu) \times G(\delta, \nu)$$

Proof. Let

$$G = G(\nu, \Delta), \qquad H = G(\mu, \Delta)$$

Then by Theorem VII.8,

$$G \cong G(\delta, \mu), \quad H \cong G(\delta, \nu)$$

Suppose that $A \in \Gamma(\mu)$, $B \in \Gamma(\nu)$. Then $(A - I)(B - I) \equiv 0$ mod $\mu\nu$, so that

$$AB \equiv A + B - I \equiv BA \bmod \mu\nu$$

Hence

$$AB \equiv BA \bmod \Delta$$

from which it follows that G and H commute elementwise. Next Theorem VII.7 implies that

$$GH = G(\delta, \Delta), \qquad G \cap H = \{1\}$$

Thus it follows that

$$GH = G \times H$$

and so

$$G(\delta, \Delta) \cong G(\delta, \mu) \times G(\delta, \nu)$$

This concludes the proof.

13. Structure theorems for $G(\mu)$

Theorem VII.9 implies our first important structure theorem for the quotient group $G(\mu)$:

Theorem VII.10. Suppose that $(\mu, \nu) = 1$. Then

$$G(\mu\nu) \cong G(\mu) \times G(\nu)$$

Proof. In Theorem VII.9, δ becomes 1 and Δ becomes $\mu\nu$.

14. Reduction to prime powers

The structure of the group $G(\mu)$ is thus reduced to the structure of $G(\pi^e)$, π prime. We have

Theorem VII.11. Let $\mu = \prod_{k=1}^{s} \pi_k^{e_k}$ be the prime-power decomposition of μ. Then

$$G(\mu) \cong G(\pi_1^{e_1}) \times G(\pi_2^{e_2}) \times \cdots \times G(\pi_s^{e_s})$$

15. The groups $G(\pi^e)$

Thus the subgroups of $G(\mu)$ [and so the congruence groups containing $\Gamma(\mu)$] are determined at least up to isomorphism by the subgroups of $G(\pi^e)$, π prime. Hence we may confine our attention to the group $G(\pi^e)$, π prime.

Since $G(\pi^e) \cong SL(n, R/(\pi^e))$, this group may be thought of as the set of matrices of Γ modulo π^e. It is somewhat simpler to consider $GL(n, R/(\pi^e))$ first, which may be regarded as the set of matrices A of R_n such that the elements of A lie in some complete set of residues modulo π^e, and $(\det(A), \pi^e) = 1$. Since π is a prime, the latter condition is entirely equivalent to $(\det(A), \pi) = 1$. Thus we may write

$$A = A_0 + \pi A_1 \tag{9}$$

where the elements of A_0 lie in some prescribed complete set of residues modulo π, the elements of A_1 lie in some prescribed complete set of residues modulo π^{e-1}, and the matrix A_0 must satisfy $(\det(A_0), \pi) = 1$. In this way the structure of $GL(n, R/(\pi^e))$ is reduced to the structure of $GL(n, R/(\pi))$.

16. The orders of $GL(n, R/(\mu))$, $SL(n, R/(\mu))$

The group $GL(n, R/(\pi))$ is especially easy to discuss because the quotient ring $R/(\pi) = F$ is a field. When F is a finite field, these groups have been thoroughly studied and are the source of much of our knowledge of simple groups. In the following discussion, we work out the orders of $GL(n, F)$ and $(SL(n, F)$ when F is a finite field, and apply the results to the determination of the order of $G(\mu)$ when $N(\mu)$ is finite.

Suppose then that $F = GF(q)$, the finite field with q elements. We first count the number of nonsingular matrices of F_n.

Let A be such a matrix, and write A as the matrix of its column vectors,

$$A = [C_1, C_2, \ldots, C_n]$$

Then A is nonsingular if and only if the column vectors of A are linearly independent over F. Clearly, there are $q^n - 1$ possible choices for C_1, since only $C_1 = 0$ must be excluded. Suppose we have chosen C_1, $C_2, \ldots, C_k$. Then C_{k+1} may be chosen as any vector which does not lie in the subspace spanned by the independent vectors $C_1, C_2, \ldots, C_k$. Since this subspace contains q^k elements, and since there are q^n vectors in all, C_{k+1} may be chosen in $q^n - q^k$ ways. It follows that A may be chosen in

$$(q^n - 1)(q^n - q) \cdots (q^n - q^{n-1})$$

ways, and so we have proved

Theorem VII.12. The order of $GL(n, GF(q))$ is given by

$$q^{(n^2-n)/2}(q - 1)(q^2 - 1) \cdots (q^n - 1) \tag{10}$$

Since $SL(n, GF(q))$ is of index $q - 1$ in $GL(n, GF(q))$, we have also proved

Theorem VII.13. The order of $SL(n, GF(q))$ is given by

$$q^{(n^2-n)/2}(q^2 - 1)(q^3 - 1) \cdots (q^n - 1) \tag{11}$$

Now suppose that $N(\pi^e)$ is finite. Then because of (9), the order of $GL(n, R/(\pi^e))$ is the order of $GL(n, R/(\pi))$ multiplied by $N(\pi^{e-1})^{n^2}$. Since $R/(\pi)$ is the finite field with $N(\pi)$ elements, formula (10) implies

Theorem VII.14. Suppose that $N(\pi^e)$ is finite, and put $q = N(\pi)$. Then the order of $GL(n, R/(\pi^e))$ is given by

$$q^{en^2}\left(1 - \frac{1}{q}\right)\left(1 - \frac{1}{q^2}\right) \cdots \left(1 - \frac{1}{q^n}\right) \tag{12}$$

Next, let A be any element of $GL(n, R/(\pi^e))$. The determinant of A may take any one of

$$N(\pi)^e(1 - 1/N(\pi))$$

values, since it is relatively prime to π and lies in a complete set of residues modulo π^e. It follows that

Theorem VII.15. Suppose that $N(\pi^e)$ is finite, and put $q = N(\pi)$. Then the order of $G(\pi^e)$ is given by

$$q^{e(n^2-1)}\left(1 - \frac{1}{q^2}\right)\left(1 - \frac{1}{q^3}\right) \cdots \left(1 - \frac{1}{q^n}\right) \tag{13}$$

Combining this result with Theorem VII.11, we have

Theorem VII.16. Suppose that $N(\mu)$ is finite. Then the order of $G(\mu)$ is given by

$$N(\mu)^{n^2-1} \prod_{\pi \mid \mu} \prod_{k=2}^{n} \left(1 - \frac{1}{N(\pi)^k}\right) \tag{14}$$

17. The groups $\Gamma^*(\mu)$

Before leaving the principal congruence groups, we mention the groups $\Gamma^*(\mu)$, consisting of all matrices of Γ congruent to a scalar matrix modulo μ. It is clear that $\Gamma^*(\mu)$ is also a normal subgroup of Γ and that $\Gamma^*(\mu)/\Gamma(\mu)$ is isomorphic to the group of elements λ of R such that $\lambda^n \equiv 1 \bmod \mu$, and which lie in some complete set of residues modulo μ. Since any such λ must be relatively prime to μ and hence must satisfy $\lambda^{\varphi(\mu)} \equiv 1 \bmod \mu$, it must also be true that

$$\lambda^d \equiv 1 \bmod \mu, \qquad d = (n, \varphi(\mu)) \tag{15}$$

The order of $\Gamma^*(\mu)/\Gamma(\mu)$ is just the number of solutions of (15) which are incongruent modulo μ. Thus if $(n, \varphi(\mu)) = 1$, the only solution of (15) modulo μ is $\lambda = 1$, and so $\Gamma^*(\mu) = \Gamma(\mu)$ in this instance.

It was shown [43, 44] that when $R = Z$, the only normal subgroups of Γ contraining $\Gamma^*(\mu)$ are the groups $\Gamma^*(\delta)$, where $\delta \mid \mu$, provided that $(\mu, 6) = 1$ for $n = 2$ and $(\mu, 2) = 1$ for $n > 2$. The restriction $(\mu, 6) = 1$ was removed [30], where all normal subgroups of Γ containing $\Gamma^*(\mu)$ were determined.

18. The congruence groups $\Gamma_{r,s}(\mu)$

We now introduce a class of congruence groups which are of particular importance in the classical case $n = 2$, $R = Z$. Let r, s be fixed, positive integers, $r + s = n$. Let M be any element of Γ, and write

$$M = \begin{bmatrix} A & B \\ C & D \end{bmatrix}$$

where A is $r \times r$, B is $r \times s$, C is $s \times r$, and D is $s \times s$. Let μ be any nonzero element of R. Then we define $\Gamma_{r,s}(\mu)$ as the subgroup of Γ consisting of all matrices M for which

$$C \equiv 0 \bmod \mu$$

These groups are not in general normal subgroups of Γ. In view of Theorem VII.6, the quotient group $\Gamma_{r,s}(\mu)/\Gamma(\mu)$ may be regarded as the set of matrices $\left[\begin{smallmatrix} A & B \\ 0 & D \end{smallmatrix}\right]$ of R_n such that the elements of A, B, D lie in some prescribed complete set of residues modulo μ, and $\det(A)\cdot\det(D) \equiv 1 \bmod \mu$.

19. The order of $\Gamma_{r,s}(\mu)/\Gamma(\mu)$

We first work out the order of $\Gamma_{r,s}(\mu)/\Gamma(\mu)$ when $N(\mu)$ is finite. In view of Theorem VII.11, it is only necessary to consider the case $\mu = \pi^e$, π prime. Set $q = N(\pi)$. Then the number of choices for A, if $\det(A) = \alpha$ is fixed, is the same as the number of elements of $GL(r, R/(\pi^e))$ having determinant α; and by Theorem VII.14 and the remarks that follows it, this is just

$$q^{e(r^2-1)}\left(1 - \frac{1}{q^2}\right)\left(1 - \frac{1}{q^3}\right) \cdots \left(1 - \frac{1}{q^r}\right) \tag{16}$$

Similarly, the number of choices for D is the same as the number of elements of $GL(s, R/(\pi^e))$ having determinant δ, where $\alpha\delta \equiv 1 \bmod \mu$, and this is just

$$q^{e(s^2-1)}\left(1 - \frac{1}{q^2}\right)\left(1 - \frac{1}{q^3}\right) \cdots \left(1 - \frac{1}{q^s}\right) \tag{17}$$

Finally, the number of choices for B is just

$$q^{ers} \tag{18}$$

Thus the total number of choices is the product of (16), (17), and

(18), multiplied by the total number of choices for α, which is $q^e(1 - 1/q)$, since α runs over all elements lying in a complete set of residues modulo π^e and relatively prime to π^e. The order of $\Gamma_{r,s}(\pi^e)/\Gamma(\pi^e)$ is then

$$q^{e(r^2+rs+s^2-1)} \frac{Q_r Q_s}{1-q}$$

where we have put

$$Q_n = \left(1 - \frac{1}{q}\right)\left(1 - \frac{1}{q^2}\right) \cdots \left(1 - \frac{1}{q^n}\right)$$

It follows that the order of $\Gamma_{r,s}(\mu)/\Gamma(\mu)$ is given by

$$N(\mu)^{r^2+rs+s^2-1} \prod_{\pi|\mu} \frac{Q_r Q_s}{1 - N(\pi)}$$

where

$$Q_n = \prod_{k=1}^{n} \left(1 - \frac{1}{N(\pi)^k}\right) \tag{19}$$

20. The index of $\Gamma_{r,s}(\mu)$

If we now make use of (14), we get

Theorem VII.17. Suppose that $N(\mu)$ is finite. Define Q_n by (19) for every prime π dividing μ. Then the index of $\Gamma_{r,s}(\mu)$ in Γ is given by

$$(\Gamma : \Gamma_{r,s}(\mu)) = N(\mu)^{rs} \prod_{\pi|\mu} \frac{Q_n}{Q_r Q_s} \tag{20}$$

For example, if $r = s = 1$, $n = 2$, and $R = Z$, (20) becomes

$$\mu \prod_{p|\mu} \left(1 + \frac{1}{p}\right)$$

which is the classical formula.

21. The structure of $\Gamma_{r,s}(\mu)$ and an important lemma

We now develop some results which give information on the structure of the groups $\Gamma_{r,s}(\mu)$. First it is clear that if μ, ν are arbitrary nonzero elements of R, then

$$\Gamma_{r,s}(\mu) \cap \Gamma_{r,s}(\nu) = \Gamma_{r,s}([\mu, \nu]) \tag{21}$$

where $[\mu, \nu]$ denotes the least common multiple of μ and ν. Next, let

$$M = \begin{bmatrix} A & B \\ C & D \end{bmatrix}$$

be any element of Γ, partitioned so that A is $r \times r$, B is $r \times s$, C is $s \times r$, D is $s \times s$. The lemma that follows was proved [53] for the case $R = Z$, and finds numerous applications.

Lemma VII.3. Let μ be any given nonzero element of R. Then an $r \times s$ matrix X over R may be determined such that

$$(\det(A + XC), \mu) = 1$$

Proof. It is sufficient to show that for every prime π, there is an $r \times s$ matrix X_π over R such that $(\det(A + X_\pi C), \pi) = 1$. For we may then find an $r \times s$ matris X over R such that

$$X \equiv X_\pi \bmod \pi$$

for each of the distinct primes π dividing μ. Since

$$\det(A + XC) \equiv \det(A + X_\pi C) \bmod \pi$$

it then follows that $(\det(A + XC), \mu) = 1$.

Now let π be a fixed prime, and let $A_1, A_2, \ldots, A_r$ denote the rows of A and $C_1, C_2, \ldots, C_s$ the rows of C. Since the rows of $X_\pi C$ are linear combinations of the rows of C, we need only show that there exist linear combinations

$$E_i = \sum_{j=1}^{s} x_{ij} C_j, \qquad 1 \leq i \leq r, \qquad x_{ij} \in R$$

such that the vectors

$$A_i + E_i, \qquad 1 \leq i \leq r$$

are linearly independent modulo π [equivalently, are linearly independent over the field $R/(\pi)$, if the matrix M is regarded as an element of $SL(n, R/(\pi))$].

Since M is of determinant 1, the rank of the $n \times r$ matrix $\left[\begin{smallmatrix} A \\ C \end{smallmatrix}\right]$ modulo π must be r, and so the set of vectors $\{A_1, A_2, \ldots, A_r, C_1, C_2, \ldots, C_s\}$ contains exactly r linearly independent vectors modulo π. Suppose that the rows of A contain exactly r' linearly independent vectors modulo π, where $r' \leq r$; and for simplicity of notation, suppose that these are $A_1, A_2, \ldots, A_{r'}$. Then each A_k $(r' < k \leq r)$ is a linear combination modulo π of $A_1, A_2, \ldots, A_{r'}$. Further, there exist $r - r'$ vectors $C_1^*, C_2^*, \ldots, C_{r-r'}^*$ among $C_1, C_2, \ldots, C_s$ such that the set $\{A_1, A_2, \ldots, A_{r'}, C_1^*, C_2^*, \ldots, C_{r-r'}^*\}$ is linearly independent modulo π. Then we need only choose

$$E_1 = E_2 = \cdots = E_{r'} = 0$$
$$E_{r'+1} = C_1^*, E_{r'+2} = C_2^*, \ldots, E_r = C_{r-r'}^*$$

to achieve the desired result. This completes the proof.

22. The case $r = s = 1$

The special case when $r = s = 1$ is worth noting, since it provides a substitute for Dirichlet's theorem, which is not available in arbitrary principal ideal rings. This says that if α, γ are relatively prime elements of R, then the "arithmetic progression" $\alpha + x\gamma$ always contains a term relatively prime to any preassigned nonzero element μ of R.

23. Coset representatives

Now if X is any $r \times s$ matrix over R and Y any $s \times r$ matrix over R, define

$$S(X) = \begin{bmatrix} I & X \\ 0 & I \end{bmatrix}, \qquad W(Y) = \begin{bmatrix} I & 0 \\ Y & I \end{bmatrix}$$

The next lemma gives information on the structure of the coset representatives of $\mathbf{\Gamma}_{r,s}(\mu)$ modulo $\mathbf{\Gamma}_{r,s}(\mu\nu)$.

Lemma VII.4. Let M be any element of $\mathbf{\Gamma}_{r,s}(\mu)$. Then X and Y may be determined so that

$$W(\mu Y)\, S(X)\, M \in \mathbf{\Gamma}_{r,s}(\mu\nu) \tag{22}$$

The entries of X and Y may be chosen modulo ν. It follows that the set of products

$$S(X)\, W(\mu Y)$$

as X runs over all $r \times s$ matrices modulo ν and Y runs over all $s \times r$ matrices modulo ν, contains a full set of left coset representatives of $\mathbf{\Gamma}_{r,s}(\mu)$ modulo $\mathbf{\Gamma}_{r,s}(\mu\nu)$.

Proof. Set

$$M = \begin{bmatrix} A & B \\ \mu C & D \end{bmatrix}$$

By Lemma VII.4, we can determine X modulo ν so that $(\det(A + \mu XC), \nu) = 1$. Set $A_0 = A + \mu XC$. Then

$$S(X)M = \begin{bmatrix} A_0 & * \\ \mu C & * \end{bmatrix}$$

and

$$W(\mu Y)\, S(X)\, M = \begin{bmatrix} * & * \\ \mu(YA_0 + C) & * \end{bmatrix}$$

In order for (22) to hold, we need only show that Y modulo ν may be determined so that

$$YA_0 + C \equiv 0 \bmod \nu$$

Now $(\det(A_0), \nu) = 1$, so that an element α of R exists such that

$$\alpha \det(A_0) \equiv 1 \bmod \nu$$

Set

(23) $$Y \equiv -\alpha C A_0^{\mathrm{adj}} \bmod \nu$$

Then

$$\begin{aligned} YA_0 &\equiv -\alpha C A_0^{\mathrm{adj}} A_0 \\ &\equiv -\alpha \det(A_0) C \\ &\equiv -C \bmod \nu \end{aligned}$$

as desired.

24. An inclusion theorem for $\boldsymbol{\Gamma}_{r,s}(\boldsymbol{\mu})$

We are now prepared to prove an interesting inclusion theorem for the groups $\Gamma_{r,s}(\mu)$ which was proved for $R = Z$ and all r, s in [38] and [53]. In order to generalize the theorem for any principal ideal ring R, we must exclude $r = s = 1$. For simplicity, we assume that both r and s are ≥ 2.

Theorem VII.18. Let $\mathbf{\Gamma}_n = SL(n, R)$. Let μ, ν be nonzero elements of R. Suppose that $r \geq 2$, $s \geq 2$. Then the only subgroups G of $\mathbf{\Gamma}_n$ such that

$$\mathbf{\Gamma}_{r,s}(\mu\nu) \subset G \subset \mathbf{\Gamma}_{r,s}(\mu) \tag{24}$$

are the groups $G = \mathbf{\Gamma}_{r,s}(\mu\delta)$, where $\delta \mid \nu$.

Proof. The proof is by induction on the total number $\Omega(\nu)$ of prime factors of ν.

If $\Omega(\nu) = 0$, then ν is a unit, and (24) implies that $G = \mathbf{\Gamma}_{r,s}(\mu)$.

Suppose the theorem proved for all μ, and for all ν such that $\Omega(\nu) < k$. Suppose ν is such that $\Omega(\nu) = k$. Put $\nu = \rho\nu'$, where ρ is any divisor of ν not a unit. By (24), we have

$$\mathbf{\Gamma}_{r,s}(\mu\rho\nu') \subset G \subset \mathbf{\Gamma}_{r,s}(\mu)$$

Intersecting by $\mathbf{\Gamma}_{r,s}(\mu\rho)$, and using (21), we get

$$\mathbf{\Gamma}_{r,s}(\mu\rho\nu') \subset G \cap \mathbf{\Gamma}_{r,s}(\mu\rho) \subset \mathbf{\Gamma}_{r,s}(\mu\rho)$$

Applying the induction hypothesis to ν' [as we may, since $\Omega(\nu') < k$] and with $\mu\rho$ replacing μ, we obtain

$$G \cap \mathbf{\Gamma}_{r,s}(\mu\rho) = \mathbf{\Gamma}_{r,s}(\mu\rho\sigma), \qquad \text{where} \quad \sigma \mid \nu'$$

Hence

$$\mathbf{\Gamma}_{r,s}(\mu\rho\sigma) \subset G \subset \mathbf{\Gamma}_{r,s}(\mu)$$

There are two cases to consider.

Case 1. For some ρ, σ and ν' are not associates.
Case 2. For every ρ, σ and ν' are associates.

In Case 1 we apply the induction hypothesis to $\rho\sigma$ [as we may, since $\Omega(\rho\sigma) < k$] to obtain $G = \mathbf{\Gamma}_{r,s}(\mu\delta)$, where $\delta \mid \rho\sigma \mid \nu$.

In Case 2, we have that for every divisor ρ of ν not a unit, $G \cap \mathbf{\Gamma}_{r,s}(\mu\rho) = \mathbf{\Gamma}_{r,s}(\mu\nu)$.

Suppose now that $G \neq \mathbf{\Gamma}_{r,s}(\mu\nu)$. The above then shows that there exists a matrix

$$M = \begin{bmatrix} A & B \\ \mu C & D \end{bmatrix} \in G$$

such that $C \not\equiv 0 \bmod \rho$ for any divisor ρ of ν, ρ not a unit. Choose X, Y as in Lemma VII.4, and use the fact that $S(X)$ belongs to G. Then we see that $W(\mu Y)$ belongs to G, where Y is chosen by use of (23). Hence also $Y \not\equiv 0 \bmod \rho$ for any divisor ρ of ν, ρ not a unit.

Call an $s \times r$ matrix T *permissible* if $W(\mu T) \in G$. We have shown the existence of a permissible matrix Y such that $Y \not\equiv 0 \bmod \rho$ for any divisor ρ of ν, ρ not a unit. We shall use this to deduce that every $s \times r$ matrix is permissible. Since already $S(X) \in G$ for all $r \times s$ matrices X over R, it will then follows from Lemma VII.4 that $G = \mathbf{\Gamma}_{r,s}(\mu)$, and the theorem will be proved.

Now we have

$$W(\mu T_1) \cdot W(\mu T_2) = W(\mu(T_1 + T_2))$$

and

$$\begin{bmatrix} V^{-1} & 0 \\ 0 & U \end{bmatrix} W(\mu T) \begin{bmatrix} V & 0 \\ 0 & U^{-1} \end{bmatrix} = W(\mu UTV)$$

where $U \in \mathbf{\Gamma}_s$, $V \in \mathbf{\Gamma}_r$. Therefore if T_1 and T_2 are permissible, so is $T_1 + T_2$. If T is permissible, then so is $-T$; and if $U \in \mathbf{\Gamma}_s$, $V \in \mathbf{\Gamma}_r$, then UTV is also permissible.

Starting with the permissible Y above, set $Y_1 = UYV$, where $U \in \mathbf{\Gamma}_s$, $V \in \mathbf{\Gamma}_r$. Then Y_1 is also permissible, and may be taken in Smith form

$$Y_1 = \begin{bmatrix} h_1 & & & & \\ & h_2 & & & \\ & & \cdot & & \\ & & & \cdot & \\ & & & & \cdot & \\ & & & & & h_t \end{bmatrix}, \qquad t = \min(r, s)$$

where $h_1 \mid h_2 \mid \cdots \mid h_t$. If $(h_1, \nu) > 1$, then there is a prime $\pi \mid \nu$ such that $Y_1 \equiv 0 \bmod \pi$. Then also $Y \equiv 0 \bmod \pi$, which is impossible. Hence $(h_1, \nu) = 1$.

Now

$$Y_2 = \left(\begin{bmatrix} 0 & 1 \\ -1 & 0 \end{bmatrix} \dotplus I_{s-2} \right) Y_1 = \begin{bmatrix} 0 & h_2 & & & & \\ -h_1 & 0 & h_3 & & & \\ & & & \cdot & & \\ & & & & \cdot & \\ & & & & & \cdot \\ & & & & & & h_t \end{bmatrix}$$

and

$$Y_3 = Y_1 - Y_2 = \begin{bmatrix} h_1 & -h_2 & & & & \\ h_1 & h_2 & & & & \\ & & 0 & & & \\ & & & \cdot & & \\ & & & & \cdot & \\ & & & & & \cdot \\ & & & & & & 0 \end{bmatrix}$$

are permissible. In Y_3 subtract the second row from the first row, and then subtract the matrix so obtained form Y_3, obtaining the permissible matrix which has h_1 in the (1, 1) position, h_2 in the (1, 2) position, and 0 elsewhere. In this matrix add $-h_2/h_1$ times column 1 to column 2, thereby obtaining the permissible matrix

$$Y_4 = \begin{bmatrix} h_1 & & & & \\ & 0 & & & \\ & & \cdot & & \\ & & & \cdot & \\ & & & & \cdot \\ & & & & & 0 \end{bmatrix}$$

Let x be any element of R. Then also

$$Y_5 = Y_4\left(\begin{bmatrix} x & 1 \\ -1 & 0 \end{bmatrix} \dotplus I_{r-2}\right) = \begin{bmatrix} h_1x & h_1 & & & & \\ 0 & 0 & & & & \\ & & 0 & & & \\ & & & \cdot & & \\ & & & & \cdot & \\ & & & & & \cdot \\ & & & & & & 0 \end{bmatrix}$$

is permissible as is

$$Y_6 = Y_4\left(\begin{bmatrix} 0 & 1 \\ -1 & 0 \end{bmatrix} \dotplus I_{r-2}\right) = \begin{bmatrix} 0 & h_1 & & & & \\ 0 & 0 & & & & \\ & & 0 & & & \\ & & & \cdot & & \\ & & & & \cdot & \\ & & & & & \cdot \\ & & & & & & 0 \end{bmatrix}$$

It follows that

$$Y_7 = Y_5 - Y_6 = \begin{bmatrix} h_1 x & & & & \\ & 0 & & & \\ & & \cdot & & \\ & & & \cdot & \\ & & & & \cdot \\ & & & & & 0 \end{bmatrix}$$

is permissible. Now $h_1 x$ runs over all elements modulo ν as x does, since $(h_1, \nu) = 1$. Since also $U\, Y_7\, V$ is permissible for all $U \in \mathbf{\Gamma}_s$, $V \in \mathbf{\Gamma}_r$, we find that every matrix whose entries are all zeros except for a single entry, which may have any value modulo ν, must be permissible. Therefore all matrices are permissible, and the theorem is proved.

Generalizations and extensions of this theorem may be found in [39], [53], and [65].

25. The symplectic group $\mathbf{\Gamma} = Sp(2n, R)$

The next group we wish to consider is the symplectic group, which is of importance because of its numerous applications to number theory and the theory of modular functions of many variables, especially as developed by Siegel in [66] and in numerous other papers.

Let J denote the $(2n) \times (2n)$ matrix

$$J = \begin{bmatrix} 0 & I \\ -I & 0 \end{bmatrix}$$

where the partitioning is such that each block is $n \times n$. Thus J is a skew-symmetric matrix such that $J^2 = -I$, and is permutation congruent to the direct sum of n copies of $T = \left[\begin{smallmatrix} 0 & 1 \\ -1 & 0 \end{smallmatrix}\right]$. Then the *symplectic* group is defined as the totality of $(2n) \times (2n)$ matrices M over R such that

$$MJM^{\mathrm{T}} = J$$

(equivalently, the *automorphs* of J). We denote this group by

$$Sp(2n, R)$$

Again, in order to avoid cumbersome notation, we write $\mathbf{\Gamma} = \mathbf{\Gamma}_{2n}$ instead of $Sp(2n, R)$.

It is readily verified that M belongs to Γ if and only if M^{T} belongs to Γ. If

$$M = \begin{bmatrix} A & B \\ C & D \end{bmatrix}$$

then the $n \times n$ matrices A, B, C, D satisfy

$$AD^{\mathrm{T}} - BC^{\mathrm{T}} = I, \qquad AB^{\mathrm{T}}, CD^{\mathrm{T}} \text{ symmetric}$$

as well as

$$A^{T}D - C^{T}B = I, \qquad A^{\mathrm{T}}C, B^{T}D \text{ symmetric}$$

26. The order of Γ when R is a finite field

When R is the finite field with q elements, it was shown by Dickson in [9] that the order of Γ is

$$q^{n^2}(q^2 - 1)(q^4 - 1) \cdots (q^{2n} - 1) \tag{25}$$

27. The center of Γ

As with $SL(n, R)$, it is easy to show that the center of Γ consists of the scalar matrices; but then the special form of M implies that the only scalar matrices in Γ are $\pm I$. It is customary to consider Γ modulo its center in the applications to number theory, which is equivalent to identifying a matrix with its negative. However, we shall retain the distinction here.

28. The generators of Γ when R is a proper euclidean ring

The generators of Γ have been determined by Hua and Reiner in [18], at least when R is a proper euclidean ring. By a *proper euclidean ring*, we mean a euclidean ring R which is not a field, and for which the integer-valued function v has the additional properties that for all nonzero elements α, β of R,

$$v(\alpha) > 0, \qquad v(\alpha\beta) = v(\alpha)v(\beta)$$

In order to complete the definition we assign the value 0 to $v(0)$. It then follows that $v(\alpha) = 1$ if and only if α is a unit, and $v(\alpha) = 0$ if and only if $\alpha = 0$.

29. Substitutes for the euclidean algorithm

If R is a proper euclidean ring and $A \in R_n$, we write

$$\operatorname{abs}(A) = v(\det(A))$$

The crucial lemma proved by Hua and Reiner [18] is as follows:

Lemma VII.5. Let R be a proper euclidean ring. Let $A \in R_n$, A symmetric, and let μ be any nonzero element of R. Then either $A \equiv 0 \bmod \mu$, or else symmetric matrices B, C of R_n exist such that $A = \mu B + C$, and $0 < \operatorname{abs}(C) < v(\mu^n)$.

Proof. We may assume that $A \not\equiv 0 \bmod \mu$. The proof proceeds by induction on n. The lemma is evident for $n = 1$. Suppose that $n = 2$. Put $A = (a_{ij})$, $B = (b_{ij})$. Then

$$\det(A - \mu B) = (a_{11} - \mu b_{11})(a_{22} - \mu b_{22}) - (a_{12} - \mu b_{12})^2 \tag{26}$$

If μ divides both a_{11} and a_{22}, then it does not divide a_{12}. We can then choose B so that $a_{11} - \mu b_{11} = a_{22} - \mu b_{22} = 0$, and $0 < v(a_{12} - \mu b_{12}) < v(\mu)$. It then follows that $\operatorname{abs}(A - \mu B) < v(\mu)^2 = v(\mu^2)$. Suppose on the other hand that μ does not divide one of the diagonal elements, say a_{11}. Choose any value for b_{12}, and choose b_{11} so that $0 < v(a_{11} - \mu b_{11}) < v(\mu)$. Since (26) may be written as

$$-\mu(a_{11} - \mu b_{11})b_{22} + \cdots$$

where the terms not indicated do not depend on b_{22}, we can choose b_{22} so that

$$\begin{aligned} 0 < \operatorname{abs}(A - \mu B) &\leq v(\mu(a_{11} - \mu b_{11})) \\ &= v(\mu)v(a_{11} - \mu b_{11}) < v(\mu)v(\mu) = v(\mu^2) \end{aligned}$$

Suppose now that the result has been established for $n = r - 1$ with $r \geq 3$. Let A be a symmetric $r \times r$ matrix over R and let some element a_{ij} of A be not divisible by μ. Since $r \geq 3$, there is some diagonal element a_{kk} of A which is not in the same row or column as a_{ij}. Let A_1 be the $(r - 1) \times (r - 1)$ symmetric matrix obtained from A by deleting the kth row and kth column of A. Let B_1 be obtained from B in similar fashion. By the induction hypothesis, B_1 may be chosen symmetric so that

$$0 < \operatorname{abs}(A_1 - \mu B_1) < v(\mu^{r-1})$$

However, we have

$$\det(A - \mu B) = (a_{kk} - \mu b_{kk}) \det(A_1 - \mu B_1) + \cdots$$

where the terms not indicated do not involve b_{kk}. Choose b_{lk} arbitrarily for $1 \leq l \leq r$, $l \neq k$, so that B is symmetric. Then choose b_{kk} so that

$$\begin{aligned} 0 < \operatorname{abs}(A - \mu B) &\leq v(\mu \det(A_1 - \mu B_1)) \\ &= v(\mu) \operatorname{abs}(A_1 - \mu B_1) \\ &< v(\mu) v(\mu^{r-1}) \\ &= v(\mu^r) \end{aligned}$$

Thus the desired result holds for $n = r$, and the induction is complete.

This lemma allows us to prove a result which serves as a substitute for the euclidean algorithm:

Lemma VII.6. Let R be a proper euclidean ring. Let A, B be elements of R_n such that AB^{adj} is symmetric, and $\det(B) = \mu \neq 0$. Then a symmetric matrix X of R_n exists such that either $A = XB$, or else $0 < \operatorname{abs}(A - XB) < \operatorname{abs}(B)$.

Proof. By Lemma VII.5, there is a symmetric matrix X of R_n such that either $AB^{\text{adj}} = \mu X$ or else

(27) $$0 < \operatorname{abs}(AB^{\text{adj}} - \mu X) < v(\mu^n)$$

Suppose the latter. Then

$$\begin{aligned} AB^{\text{adj}} - \mu X &= AB^{\text{adj}} - XBB^{\text{adj}} \\ &= (A - XB)B^{\text{adj}} \end{aligned}$$

Thus $\det(AB^{\text{adj}} - \mu X) = \det(A - XB) \cdot \mu^{n-1}$, and (27) implies that

$$0 < \operatorname{abs}(A - XB) < v(\mu) = \operatorname{abs}(B)$$

This completes the proof.

30. A useful lemma for any principal ideal ring

We also require the following useful result, which is true for any principal ideal ring:

Lemma VII.7. Let R be any principal ideal ring. Let μ be any nonzero element of R. Let

$$M = \begin{bmatrix} A & B \\ C & D \end{bmatrix}$$

be any element of Γ. Then a symmetric matrix X of R_n exists such that

$$(\det(A + XC), \mu) = 1$$

Proof. As in the proof of Lemma VII.3, it suffices to show that for each prime π there is a symmetric matrix X_π such that $(\det(A + X_\pi C), \pi) = 1$, since the construction of X by the Chinese remainder theorem preserves the symmetry. Let U, V be any elements of $SL(n, R)$. Then

$$\begin{bmatrix} U & 0 \\ 0 & (U^{\mathrm{T}})^{-1} \end{bmatrix}, \quad \begin{bmatrix} V & 0 \\ 0 & (V^{\mathrm{T}})^{-1} \end{bmatrix}$$

belong to Γ, and

$$\begin{bmatrix} U & 0 \\ 0 & (U^{\mathrm{T}})^{-1} \end{bmatrix} M \begin{bmatrix} V & 0 \\ 0 & (V^{\mathrm{T}})^{-1} \end{bmatrix} = \begin{bmatrix} A_1 & B_1 \\ C_1 & D_1 \end{bmatrix}$$

where $A_1 = UAV$, $C_1 = (U^{\mathrm{T}})^{-1}CV$. Set $Y_\pi = UX_\pi U^{\mathrm{T}}$. Then

$$A_1 + Y_\pi C_1 = U(A + X_\pi C)V$$

Hence we need only find a symmetric matrix Y_π such that $(\det(A_1 + Y_\pi C_1), \pi) = 1$.

By proper choice of $U, V \in SL(n, R)$, we may assume that A_1 is diagonal. If $A_1 \equiv 0 \bmod \pi$, then the choice $Y_\pi = I$ will do. If $A_1 \not\equiv 0 \bmod \pi$, write

$$A_1 \equiv \begin{bmatrix} E & 0 \\ 0 & 0 \end{bmatrix} \bmod \pi$$

where E is diagonal and nonsingular modulo π. Partition C_1 in similar fashion:

$$C_1 = \begin{bmatrix} C_{11} & C_{12} \\ C_{21} & C_{22} \end{bmatrix}$$

The symmetry of $A_1{}^{\mathrm{T}}C_1$ shows that $C_{12} \equiv 0 \bmod \pi$. Hence

$$\begin{bmatrix} A_1 \\ C_1 \end{bmatrix} \equiv \begin{bmatrix} E & 0 \\ 0 & 0 \\ C_{11} & 0 \\ C_{21} & C_{22} \end{bmatrix} \bmod \pi$$

whence $(\det(C_{22}), \pi) = 1$. Thus if

$$Y_\pi = \begin{bmatrix} 0 & 0 \\ 0 & I \end{bmatrix}$$

then

$$A_1 + Y_\pi C_1 \equiv \begin{bmatrix} E & 0 \\ C_{21} & C_{22} \end{bmatrix} \bmod \pi$$

which is nonsingular modulo π. This completes the proof.

The discussion now proceeds as follows: Define

$$S(X) = \begin{bmatrix} I & X \\ 0 & I \end{bmatrix}, \qquad X \in R_n, \qquad X \text{ symmetric},$$

$$D(U) = \begin{bmatrix} U & 0 \\ 0 & (U^{\mathrm{T}})^{-1} \end{bmatrix}, \qquad U \in GL(n, R)$$

Then $S(X)$, $D(U)$ belong to Γ. Let $M = [\begin{smallmatrix} A & B \\ C & D \end{smallmatrix}]$ be any element of Γ. By Lemma VII.7, a symmetric matrix $X_0 \in R_n$ may be determined so that $\det(A + X_0 C)$ is prime to any preassigned nonzero element of R, and hence certainly nonsingular. Put

$$M_1 = JS(X_0)M = \begin{bmatrix} A_1 & B_1 \\ C_1 & D_1 \end{bmatrix}$$

Then C_1 is nonsingular. Now the symmetry of $A_1{}^{\mathrm{T}} C_1$ implies the symmetry of $A_1 C_1^{\mathrm{adj}}$, and so by Lemma VII.6, a symmetric matrix X_1 may be determined so that either $A_1 = X_1 C_1$, or else $0 < \operatorname{abs}(A_1 - X_1 C_1) < \operatorname{abs}(C_1)$. In the latter case, repeat the entire procedure with the matrix

$$M_2 = JS(-X_1)M_1 = \begin{bmatrix} A_2 & B_2 \\ C_2 & D_2 \end{bmatrix}$$

in which $0 < \operatorname{abs}(C_2) < \operatorname{abs}(C_1)$. Eventually this process must terminate, and a matrix

$$M_k = \begin{bmatrix} A_k & B_k \\ C_k & D_k \end{bmatrix}$$

is reached for which a symmetric matrix X_k exists such that $A_k = X_k C_k$. But then

$$M_{k+1} = JS(-X_k)M_k = \begin{bmatrix} A_{k+1} & B_{k+1} \\ 0 & D_{k+1} \end{bmatrix}$$

Since $D_{k+1} = (A_{k+1}^T)^{-1}$, we get

$$M_{k+1} = D(A_{k+1})\, S\,(A_{k+1}^{-1}B_{k+1})$$

We have thus proved

Theorem VII.19. Let R be a proper euclidean ring. Then the symplectic group $\Gamma = Sp(2n, R)$ is generated by the matrices

$$S(X) = \begin{bmatrix} I & X \\ 0 & I \end{bmatrix}, \qquad X \text{ symmetric, } X \in R_n,$$

$$D(U) = \begin{bmatrix} U & 0 \\ 0 & (U^{\mathrm{T}})^{-1} \end{bmatrix}, \qquad U \in GL(n, R),$$

$$J = \begin{bmatrix} 0 & I \\ -I & 0 \end{bmatrix}$$

There remains the problem of reducing this list. We will not go into the details of this question, but mention only that Hua and Reiner showed that for $R = Z$, Γ may be generated by two elements when $n = 1$ and four elements when $n > 1$, which are given explicitly.

31. Principal congruence subgroups of Γ and an isomorphism

In analogy with the discussion given for $SL(n, R)$, we now define the *principal congruence subgroup* of level μ of $\Gamma = Sp(2n, R)$, μ a nonzero element of the principal ideal ring R, as the set of all matrices A of Γ such that

$$A \equiv I \bmod \mu$$

Again, we denote this normal subgroup by $\Gamma(\mu)$, the quotient group $\Gamma/\Gamma(\mu)$ by $G(\mu)$, and the quotient group $\Gamma(\alpha)/\Gamma(\beta)$ by $G(\alpha, \beta)$, where α, β are elements of R such that α/β. Many of the results proved previously for $SL(n, R)$ go over unchanged and with similar proofs for $Sp(2n, R)$. Thus we have

Theorem VII.20. If $\Gamma = Sp(2n, r)$, then we have the isomorphism

$$G(\mu) = \Gamma/\Gamma(\mu) \cong SP(2n, R/(\mu))$$

32. Congruence groups

As before, a *congruence group* will mean a subgroup of Γ which contains a principal congruence group. When $R = Z$ and $n > 1$,

Mennicke has shown in [33] that any subgroup of Γ of finite index is a congruence group, thereby indicating their importance.

We want to get at the structure of $G(\mu)$, for which we need the result that corresponds to Theorem VII.7 for $SL(n, R)$. This is somewhat more difficult to obtain, and motivates the discussion that follows.

33. Symmetric and symplectic matrices modulo μ

Let μ be a nonzero element of the principal ideal ring R. If A is a matrix of R_n such the $A \equiv A^{\mathrm{T}} \bmod \mu$, we say that A is *symmetric modulo* μ; and if M is a matrix of R_{2n} such that $MJM^{\mathrm{T}} \equiv J \bmod \mu$, we say that M is *symplectic modulo* μ. The results that follow were given in [55] for the case $R = Z$. To avoid certain trivial exceptions, we assume that R is not of characteristic 2 in what follows.

Lemma VII.8. Suppose that A belongs to R_n and that A is symmetric modulo μ. Then there is a symmetric matrix B of R_n such that $B \equiv A \bmod \mu$.

Proof. Put $A = A^{\mathrm{T}} + \mu E$, where E belongs to R_n. Then E is skew symmetric. Put $E = (e_{ij})$, and define

$$E_1 = \begin{bmatrix} 0 & e_{12} & e_{13} & \cdots & e_{1n} \\ 0 & 0 & e_{23} & \cdots & e_{2n} \\ \cdot & \cdot & \cdot & & \cdot \\ \cdot & \cdot & \cdot & & \cdot \\ \cdot & \cdot & \cdot & & \cdot \\ 0 & 0 & 0 & \cdots & 0 \end{bmatrix}$$

Then certainly $E = E_1 - E_1{}^{\mathrm{T}}$. Hence we may choose $B = A - \mu E_1$, since

$$\begin{aligned} B - B^{\mathrm{T}} &= A - \mu E_1 - (A^{\mathrm{T}} - \mu E_1{}^{\mathrm{T}}) \\ &= (A - A^{\mathrm{T}}) - \mu(E_1 - E_1{}^{\mathrm{T}}) \\ &= \mu E - \mu E = 0 \end{aligned}$$

Lemma VII.9. Let $M = \left[\begin{smallmatrix} A & B \\ C & D \end{smallmatrix}\right] \in R_{2n}$, and suppose that M is symplectic modulo μ. Then there is a symmetric matrix $X \in R_n$ such that

$$(\det(A + XC), \mu) = 1$$

Proof. The proof, with minor modifications, is identical with the proof of Lemma VII.7. The essential point is that $(\det(M), \mu) = 1$, since $(\det(M))^2 \equiv 1 \bmod \mu$.

Lemma VII.10. Suppose that P, Q are commuting symmetric matrices of R_n such that $M = \left[\begin{smallmatrix} P & 0 \\ 0 & Q \end{smallmatrix}\right]$ is symplectic modulo μ. Then there is a symplectic matrix N such that $M \equiv N \bmod \mu$.

Proof. M is symplectic modulo μ if and only if $PQ \equiv I \bmod \mu$. Put $PQ = I - \mu E$, where $E \in R_n$, and is symmetric and commutes with both P and Q. Then it is easily verified that the matrix

$$N = \begin{bmatrix} P + \mu EP & -\mu E \\ \mu E & Q \end{bmatrix}$$

is symplectic, and is certainly congruent to M modulo μ.

34. Completing symplectic matrices modulo μ to symplectic matrices

We now prove the analog of (b) of Section II.21.

Theorem VII.21. Suppose that $M = \left[\begin{smallmatrix} A & B \\ C & D \end{smallmatrix}\right] \in R_{2n}$, and that M is symplectic modulo μ. Then there is a symplectic matrix N such that $M \equiv N \bmod \mu$.

Proof. By Lemma VII.9, there is a symmetric matrix $X \in R_n$ such that $(\det(A + XC), \mu) = 1$. Put

$$M_1 = \begin{bmatrix} I & X \\ 0 & I \end{bmatrix} M = \begin{bmatrix} A + XC & B + XD \\ C & D \end{bmatrix} = \begin{bmatrix} A_1 & B_1 \\ C & D \end{bmatrix}$$

Then M_1 is also symplectic modulo μ, and $(\det(A_1), \mu) = 1$. Define α by $\alpha \det(A_1) \equiv 1 \bmod \mu$. Then $-\alpha C A_1^{\mathrm{adj}}$ is symmetric modulo μ. By Lemma VII.8, there is a symmetric matrix $Y \in R_n$ such that $Y \equiv -\alpha C A_1^{\mathrm{adj}} \bmod \mu$. Put

$$M_2 = \begin{bmatrix} I & 0 \\ Y & I \end{bmatrix} M_1 = \begin{bmatrix} A_1 & B_1 \\ YA_1 + C & YB_1 + D \end{bmatrix} = \begin{bmatrix} A_1 & B_1 \\ C_1 & D_1 \end{bmatrix}$$

Then M_2 is symplectic modulo μ, and

$$C_1 = YA_1 + C \equiv -\alpha C A_1^{\mathrm{adj}} A_1 + C \equiv 0 \bmod \mu$$

Similarly, there is a symmetric matrix $W \in R_n$ such that

$$M_3 = M_2\begin{bmatrix} I & W \\ 0 & I \end{bmatrix} = \begin{bmatrix} A_1 & A_1W + B_1 \\ C_1 & C_1W + D_1 \end{bmatrix} = \begin{bmatrix} A_1 & B_2 \\ C_1 & D_2 \end{bmatrix}$$

where $B_2 \equiv 0 \bmod \mu$. Thus M_3 is symplectic modulo μ and

$$M_3 \equiv \begin{bmatrix} A_1 & 0 \\ 0 & D_2 \end{bmatrix} \bmod \mu$$

Now determine matrices $U, V \in GL(n, R)$ such that $P = UA_1V$ is diagonal. Put

$$\begin{aligned} M_4 &= \begin{bmatrix} U & 0 \\ 0 & (U^{\mathrm{T}})^{-1} \end{bmatrix} M_3 \begin{bmatrix} V & 0 \\ 0 & (V^{\mathrm{T}})^{-1} \end{bmatrix} \\ &\equiv \begin{bmatrix} UA_1V & 0 \\ 0 & (U^{\mathrm{T}})^{-1}D_2(V^{\mathrm{T}})^{-1} \end{bmatrix} \bmod \mu \end{aligned}$$

Then M_4 is symplectic modulo μ and $M_4 \equiv \left[\begin{smallmatrix} P & 0 \\ 0 & Q \end{smallmatrix}\right] \bmod \mu$, where P is diagonal. Since $PQ^{\mathrm{T}} \equiv I \bmod \mu$, Q is congruent modulo μ to a diagonal matrix. We have shown therefore that symplectic matrices K, L exist such that

$$M \equiv K\begin{bmatrix} P & 0 \\ 0 & Q \end{bmatrix} L \bmod \mu$$

where both P and Q are diagonal matrices. To complete the proof, it is only necessary to show that a symplectic matrix N_1 exists such that

$$N_1 \equiv \begin{bmatrix} P & 0 \\ 0 & Q \end{bmatrix} \bmod \mu;$$

for then the matrix $N = KN_1L$ is also symplectic, and

$$N \equiv K\begin{bmatrix} P & 0 \\ 0 & Q \end{bmatrix} L \equiv M \bmod \mu$$

But the existence of N_1 is guaranteed by Lemma VII.10 and so the proof is complete.

35. Various structure theorems

From now on, μ, ν denote nonzero elements of R, and

$$\delta = (\mu, \nu), \qquad \Delta = [\mu, \nu]$$

The next result is the analog of Lemma VII.1.

Lemma VII.11. Suppose that $M \in \Gamma(\delta)$. Then Y can be determined so that $Y \in \Gamma(\mu)$, and

$$Y \equiv M \bmod \nu \tag{28}$$

Proof. Since $M \in \Gamma(\delta)$, we can write $M = I + \delta N$, where $N \in R_{2n}$. Set $Y = I + \mu W$. Then $Y \equiv I \bmod \mu$, and (28) becomes

$$\mu W \equiv \delta N \bmod \nu, \qquad (\mu/\delta) W \equiv N \bmod \nu/\delta$$

Since $(\mu/\delta, \nu/\delta) = 1$, this has a solution W_0. Thus there is a Y_0 such that

$$Y_0 \equiv I \bmod \mu, \qquad Y_0 \equiv M \bmod \nu$$

Then Y_0 is symplectic modulo μ and also modulo ν, since M is symplectic. Hence Y_0 is symplectic modulo Δ. Thus Y can be determined so that $Y \equiv Y_0 \bmod \Delta$ and Y is symplectic. This Y satisfies the condition of the lemma.

Exactly as in the case of $SL(n, R)$, we now obtain the following theorems:

Theorem VII.22. The normal subgroups $\Gamma(\mu)$, $\Gamma(\nu)$ of Γ satisfy

$$\Gamma(\mu)\Gamma(\nu) = \Gamma(\delta), \qquad \Gamma(\mu) \cap \Gamma(\nu) = \Gamma(\Delta)$$

Theorem VII.23. We have the isomorphism

$$G(\delta, \mu) \cong G(\nu, \Delta)$$

Theorem VII.24. Let $\times$ represent direct product. Then

$$G(\delta, \Delta) \cong G(\delta, \mu) \times G(\delta, \nu)$$

Theorem VII.25. Suppose that $(\mu, \nu) = 1$. Then

$$G(\mu\nu) \cong G(\mu) \times G(\nu)$$

Theorem VII.26. Let $\mu = \prod_{k=1}^{s} \pi_k^{e_k}$ be the prime power decomposition of μ. Then

$$G(\mu) \cong G(\pi_1^{e_1}) \times G(\pi_2^{e_2}) \times \cdots \times G(\pi_s^{e_s})$$

36. The order of $G(\mu)$

Out next task is to determine the order of $G(\mu)$ when $N(\mu)$ is finite. Because of Theorem VII.26 we need only consider $\mu = \pi^e$, π prime; and because of Theorem VII.20, we may think of $G(\mu)$ as the set of elements M of $GL(2n, R)$ modulo π^e which satisfy

$$MJM^{\mathrm{T}} \equiv J \bmod \pi^e$$

We may suppose that $e \geq 2$, since formula (25) of Dickson [9] gives the order of $G(\pi)$. Put $q = N(\pi)$, and write

$$M = M_0 + \pi^{e-1}M_1$$

where M_0 is determined modulo π^{e-1} and M_1 modulo π. Then

$$(M_0 + \pi^{e-1}M_1)J(M_0{}^{\mathrm{T}} + \pi^{e-1}M_1{}^{\mathrm{T}}) \equiv J \bmod \pi^e,$$
$$M_0JM_0{}^{\mathrm{T}} + \pi^{e-1}(M_0JM_1{}^{\mathrm{T}} + M_1 JM_0{}^{\mathrm{T}}) \equiv J \bmod \pi^e$$

It follows that

$$M_0JM_0{}^{\mathrm{T}} \equiv J \bmod \pi^{e-1};$$

and if

$$C = \frac{1}{\pi^{e-1}}(M_0JM_0{}^{\mathrm{T}} - J),$$
$$U = M_0JM_1{}^{\mathrm{T}},$$

then

(29) $$C + U - U^{\mathrm{T}} \equiv 0 \bmod \pi$$

There are thus q^{2n^2+n} possible values for U modulo π, once M_0 has been chosen, since the elements on and above the diagonal of U may be taken arbitrarily modulo π, but then the elements below the diagonal of U are determined modulo π by (29). It follows that there are the same number of possible values for M_1 modulo π, since M_0J is non-singular modulo π. Thus the order of $G(\pi^e)$ is the order of $G(\pi^{e-1})$ multiplied by q^{2n^2+n}. Since the order of $G(\pi)$ is given by (25), we get

Theorem VII.27. Let π be a prime, and suppose that $q = N(\pi)$ is finite. Then the order of $G(\pi^e)$ is

$$q^{e(2n^2+n)}\left(1 - \frac{1}{q^2}\right)\left(1 - \frac{1}{q^4}\right) \cdots \left(1 - \frac{1}{q^{2n}}\right)$$

This result and Theorem VII.26 imply

Theorem VII.28. Suppose that $N(\mu)$ is finite. Then the order of $G(\mu)$ is given by

$$N(\mu)^{2n^2+n} \prod_{\pi|\mu} \prod_{k=1}^{n} \left(1 - \frac{1}{N(\pi)^{2k}}\right)$$

37. The groups $\Gamma_0(\mu)$

In analogy with the discussion of $SL(n, R)$, we introduce a class of congruence groups of $\Gamma = Sp(2n, R)$, defined as follows:

Let $M = \left[\begin{smallmatrix} A & B \\ C & D \end{smallmatrix}\right]$ be any element of Γ. Let μ be any nonzero element of R. Then we define $\Gamma_0(\mu)$ as the totality of matrices M for which $C \equiv 0 \bmod \mu$.

Because of Theorem VII.10 we may regard $\Gamma_0(\mu)/\Gamma(\mu)$ as the set of matrices M modulo μ such that

$$\begin{gathered} C \equiv 0 \bmod \mu, \\ AD^{\mathrm{T}} \equiv I \bmod \mu, \\ AB^{\mathrm{T}} \text{ symmetric modulo } \mu \end{gathered}$$

Put $AB^{\mathrm{T}} = S$, S symmetric modulo μ. Then if A is chosen as any element of $GL(n, R/(\mu))$, B and D are determined modulo μ by S and A. It follows that if $N(\mu)$ is finite then the order of $\Gamma_0(\mu)/\Gamma(\mu)$ is just

$$N(\mu)^{(n^2+n)/2} \cdot N(\mu)^{n^2} \prod_{\pi|\mu} \prod_{k=1}^{n} \left(1 - \frac{1}{N(\pi)^{k}}\right) \tag{30}$$

where the factor $N(\mu)^{(n^2+n)/2}$ represents the number of symmetric matrices modulo μ, and the other factor represents the order of $GL(n, R/(\mu))$, given by (12) for μ a prime power.

Combining this with Theorem VII.28, we obtain

Theorem VII.29. Suppose that $N(\mu)$ is finite. Then the index of $\Gamma_0(\mu)$ in Γ is given by

$$(\Gamma : \Gamma_0(\mu)) = N(\mu)^{(n^2+n)/2} \prod_{\pi|\mu} \prod_{k=1}^{n} (1 + 1/N(\pi)^k) \tag{31}$$

Finally, we mention briefly that if μ, ν are arbitrary nonzero elements of R, then a complete set of left coset representatives for $\Gamma_0(\mu)$ modulo

$\Gamma_0(\mu\nu)$ is contained among the matrices

$$S(X)\, W(\mu Y)$$

where

$$S(X) = \begin{bmatrix} I & X \\ 0 & I \end{bmatrix}, \qquad X \in R_n, \quad X \text{ symmetric,}$$

$$W(\mu Y) = \begin{bmatrix} I & 0 \\ \mu Y & I \end{bmatrix}, \qquad Y \in R_n, \quad Y \text{ symmetric}$$

and X, Y may be chosen modulo ν. On the basis of this remark and the previous results about Γ, it is possible to prove analogues of Theorem VII.18 which hold for any principal ideal ring if $n > 1$, and hold for $n = 1$ as well when $R = Z$.

EXERCISES AND PROBLEMS

1. Let $\Gamma = SL(n, Z)$, $\Gamma_{a,b} = \Gamma(p^a)/\Gamma(p^{a+b})$, where p is an odd prime, $a > 0, b \geq 0$. Show that if $b \leq a$ then $\Gamma_{a,b}$ is abelian; and if $b > a$ then the center of $\Gamma_{a,b}$ is of order p^{an^2+b}. Hence conclude that no two of the groups $\Gamma_{a,b}$ are isomorphic for b fixed, $1 \leq a \leq b - 1$, although they are of the same order.

Hint: Consider $\Gamma_{a,b}$ as the set $I + p^a A$, A modulo p^b. Show that if $b > a$, the center consists of all matrices $I + p^a X$, X modulo p^b, X scalar modulo p^{b-a}.

2. Show that $\Gamma_{1,1}$ is of order p^{n^2-1} and abelian of type $(p, p, \ldots, p)$.

Hint: Put

$$X_{ij} = \begin{cases} I + pE_{ij}, & 1 \leq i, j \leq n, \quad i \neq j, \\ I + p(E_{ii} + E_{i,i+1} - E_{i+1,i} - E_{i+1,i+1}), & 1 \leq i \leq n-1, \\ & i = j \end{cases}$$

Show that $\Gamma_{1,1}$ is the direct product of its cyclic subgroups $\{X_{ij}\}$.

3. Show that the elements of $Sp(2n, Z)$ all have determinant 1.

Chapter VIII

The Classical Modular Group and Related Groups

1. Definition of Γ

The group of linear fractional transformations

$$\tau' = \frac{a\tau + b}{c\tau + d} \tag{1}$$

where a, b, c, d run over all integers such that $ad - bc = 1$, is known as the *classical modular group* and is denoted by Γ. Thus Γ is identical with $PSL(2, Z) = SL(2, Z)/\{\pm I\}$, where $\{\pm I\}$ is the center of $SL(2, Z)$. Many of the great advances in analytic number theory have come through a detailed study of this group and its subgroups. The connection arises via the elliptic modular functions and forms, such as the Dedekind eta function or the theta functions. These functions are invariant or semi invariant with respect to the linear transformations (1) and are the generating functions of numerous number-theoretic functions of great interest, such as the partition function, and the function that counts the number of representations of an integer by some integral positive definite quadratic form.

The group Γ can be studied geometrically, in terms of its action on the upper τ half-plane under the transformations (1), or purely group theoretically; and there is much to be gained by combining the two approaches. If the first approach is taken, the subject can be generalized to the vast theory of discrete groups, which includes the study of the Fuchsian groups and brings up questions of a topological nature, such as the determination of the automorphisms of a Riemann surface. The second approach is simpler, and leads directly to many of the important results about Γ.

At the outset it will be convenient to think of Γ as $SL(2, Z)$ modulo its center. In order to avoid confusion, elements of Γ will be denoted by lower case italic letters and elements of $SL(2, Z)$ by upper case italic letters.

2. Generators of Γ

According to the discussion of $SL(n, Z)$ given in Chapter VII, we know that $SL(2, Z)$ may be generated by the matrices

$$S = \begin{bmatrix} 1 & 1 \\ 0 & 1 \end{bmatrix}, \qquad T = \begin{bmatrix} 0 & 1 \\ -1 & 0 \end{bmatrix} \tag{2}$$

The matrix

$$U = TS = \begin{bmatrix} 0 & 1 \\ -1 & -1 \end{bmatrix} \tag{3}$$

is basic to the theory, and $SL(2, Z)$ may equally well be generated by T and U. Notice that

$$T^2 = -I, \qquad U^3 = I \tag{4}$$

3. Proof that Γ is a free product

The most important fact about Γ is undoubtedly the following:

Theorem VIII.1. The group Γ is the free product of a cyclic group $\{x\}$ of order 2 and a cyclic group $\{y\}$ of order 3:

$$\Gamma = \{x\} * \{y\}$$

Here x is the element of Γ which corresponds to the element T of

$SL(2, Z)$, any y is the element of Γ which corresponds to the element U of $SL(2, Z)$.

Proof. Consider any element A of $SL(2, Z)$. Because of (4), A may be written as

$$A = \pm U^a T U^{e_1} T U^{e_2} \cdots T U^{e_n} T^b$$

where a, b are either 0 or 1, and $e_i = 1$ or 2, $1 \leq i \leq n$. To show that Γ has the required property it is only necessary to show that no word A can reduce to $\pm I$. In order to show this, it will suffice to show that no word

$$B = \pm T U^{e_1} T U^{e_2} \cdots T U^{e_n}, \qquad e_i = 1 \text{ or } 2, \quad 1 \leq i \leq n$$

can reduce to $\pm I$, since we may assume that A is cyclically reduced as a word in T and the powers of U. This is equivalent to the statement that no word of the semigroup Δ generated by TU, TU^2 can reduce to $\pm I$. We have

$$TU = \begin{bmatrix} -1 & -1 \\ 0 & -1 \end{bmatrix} = -\begin{bmatrix} 1 & 1 \\ 0 & 1 \end{bmatrix},$$

$$TU^2 = \begin{bmatrix} 1 & 0 \\ 1 & 1 \end{bmatrix}$$

Thus if $C = \left[\begin{smallmatrix} a & b \\ c & d \end{smallmatrix}\right]$ belongs to Δ, then a, b, c, d are all of the same sign. Furthermore

$$TUC = -\begin{bmatrix} a+c & b+d \\ c & d \end{bmatrix},$$

$$TU^2C = \begin{bmatrix} a & b \\ a+c & b+d \end{bmatrix}$$

It follows that if one of the off-diagonal elements of C is different from 0, then the same is true for TUC, TU^2C. Since TU and TU^2 have this property, every element of Δ must have this property. Thus no element of Δ is $\pm I$. This completes the proof.

4. The commutator subgroup of Γ

We put

$$z = xy \tag{5}$$

so that z corresponds to S.

We now prove

Theorem VIII.2. The commutator subgroup Γ' is of index 6 in Γ. Γ/Γ' is the cyclic group of order 6 generated by $z\Gamma'$.

Proof. In Γ/Γ', the relations are $x^2 = y^3 = 1$, $xy = yx$. Hence Γ/Γ' is the direct product of the cyclic group $\{x\}$ of order 2 and the cyclic group $\{y\}$ of order 3, and so is the cyclic group $\{xy\}$ of order 6.

Because Γ' is generated by the set of commutators

$$uvu^{-1}v^{-1}, \qquad u, v \in \Gamma$$

and because Γ is the free product $\{x\} * \{y\}$, it follows that Γ' may also be characterized as the set of words w of Γ such that

$$e(w; x) \equiv 0 \bmod 2,$$
$$e(w; y) \equiv 0 \bmod 3$$

where $e(w; x)$, $e(w; y)$ are the sums of the exponents of x, y, respectively in w. In particular this implies that neither x nor y belongs to Γ'. By a standard constructive argument, it can be shown that

$$a = xyx^{-1}y^{-1} = xyxy^2, \tag{6}$$
$$b = xy^{-1}x^{-1}y = xy^2xy \tag{7}$$

are generators of Γ'. A tedious but straightforward word argument may now be used to show that Γ' is actually the free group of rank 2 freely generated by a and b. We do not prove this here, since the next theorem, which is a consequence of the Kurosh subgroup theorem, tells us much more about Γ and its subgroups.

5. The Kurosh subgroup theorem

The Kurosh subgroup theorem states that a subgroup $H \neq \{1\}$ of a free product G is itself a free product, $H = F * \Pi * G_\alpha$, where F is either free or $\{1\}$ and each G_α is the conjugate of a subgroup of one of the free factors of G. (See [23] for a proof of this theorem.) Applied to Γ, the theorem tells us that if $H \neq \{1\}$ is a subgroup of Γ, then

$$H = F * \prod * G_\alpha \tag{8}$$

where F is either free or $\{1\}$ and each G_α is conjugate to $\{x\}$ or to $\{y\}$.

6. Criterion for a subgroup to be free

This implies

Theorem VIII.3. Let $H \neq \{1\}$ be a subgroup of Γ. Then H is free if and only if it contains no elements of finite period.

Proof. If H contains no elements of finite period then the free product $\Pi * G_\alpha$ is vacuous, and so $H = F$ is free. If H is free, then it certainly contains no elements of finite period.

7. Elements of Γ of finite period

Theorem VIII.4. The only elements of finite period of Γ are the conjugates of x, y, y^2.

Proof. Let u be an element of finite period of Γ. Then the cyclic subgroup $\{u\}$ is of the form (8) if and only if $\{u\}$ is conjugate to $\{x\}$ or to $\{y\}$, which implies the result.

Another interesting consequence is that two commuting elements of Γ must be powers of the same element of Γ.

We are going to apply Theorem VIII.3 to the normal subgroups of Γ, but first we introduce some notation and define some new groups.

8. Normal closure

Let $x_1, x_2, \ldots, x_n$ be arbitrary elements of Γ. Let $\Delta = \Delta(x_1, x_2, \ldots, x_n)$ be the intersection of all normal subgroups of Γ containing $x_1, x_2, \ldots, x_n$; that is, the *normal closure* of $x_1, x_2, \ldots, x_n$. Thus Γ/Δ is the group obtained by adding the relations $x_1 = 1, x_2 = 1, \ldots, x_n = 1$ to the relations $x^2 = y^3 = 1$ of Γ.

9. The groups Γ^n

Let Γ^n be the fully invariant subgroup of Γ generated by the nth powers of the elements of Γ. Then we shall prove

Theorem VIII.5. We have

$$\Gamma^2 = \Delta(y), \qquad \Gamma^3 = \Delta(x),$$
$$(\Gamma : \Gamma^2) = 2, \qquad (\Gamma : \Gamma^3) = 3$$

$\Gamma^2 = \{y\} * \{xyx\}$ is the free product of the cyclic groups $\{y\}, \{xyx\}$ of order 3, $\Gamma^3 = \{x\} * \{yxy^2\} * \{y^2xy\}$ is the free product of the cyclic groups $\{x\}, \{yxy^2\}, \{y^2xy\}$ of order 2. Γ^2 may be characterized as the set of words w of Γ such that $e(w; x) \equiv 0 \bmod 2$, and Γ^3 as the set of words w of Γ such that $e(w; y) \equiv 0 \bmod 3$. Consequently

$$\Gamma' = \Gamma^2 \cap \Gamma^3$$

Proof. Since $y = y^{-2} \in \Gamma^2$ and Γ^2 is a normal subgroup of $\Gamma, \Gamma^2 \supset \Delta(y)$. Now $\Gamma/\Delta(y)$ is the group with relations $x^2 = y^2 = 1, y = 1$, and so is cyclic of order 2. It follows that Γ^2 must equal either Γ or $\Delta(y)$. But x certainly does not belong to Γ^2, and so $\Gamma^2 = \Delta(y)$. A similar argument shows that $\Gamma/\Delta(x) = 3$ and that $\Gamma^3 = \Delta(x)$.

It is a routine matter to show that Γ^2 is generated by y and xyx, and that Γ^3 is generated by x, yxy^2, and y^2xy. We can show that $\Gamma^2 = \{y\} * \{xyx\}$, $\Gamma^3 = \{x\} * \{yxy^2\} * \{y^2xy\}$ in the same way that Γ was shown to be a free product; namely, by going over to the corresponding matrix representation. The details of this argument may be found in [41]. Or, we can appeal to Grusko's theorem, which bounds the sum of the ranks of the factors of a free product by the rank of the free product.

Finally, it is clear that every element w of Γ^2 satisfies $e(w; x) \equiv 0 \bmod 2$, and that every element w of Γ^3 satisfies $e(w; y) \equiv 0 \bmod 3$. Thus $\Gamma' \supset \Gamma^2 \cap \Gamma^3$. But $\Gamma/\Gamma^2, \Gamma/\Gamma^3$ are abelian groups. Hence $\Gamma^2 \supset \Gamma', \Gamma^3 \cap \Gamma'$, and so $\Gamma^2 \cap \Gamma^3 \supset \Gamma'$. Thus $\Gamma' = \Gamma^2 \cap \Gamma^3$ and the proof is complete.

Γ^2, Γ^3, are the unique normal subgroups of Γ of their respective indices. Γ^n reduces to Γ, Γ^2, or Γ^3 if n is not a multiple of 6. The group Γ^6 is of index 216 in Γ, and is a free group of rank 37. The nature of Γ^{6n}, $n > 1$, is not known, but the work of Novikov and Shafarevich on the Burnside problem implies that for sufficiently large n, Γ^{6n} is of infinite index in Γ. See [40] for a discussion of this question and proofs of the results cited above.

10. Theorem on normal subgroups

We are now prepared to state our result on the normal subgroups of Γ.

Theorem VIII.6. Apart from the three exceptions $\Gamma, \Gamma^2, \Gamma^3$, every normal subgroup of Γ is free.

Proof. Let G be a normal subgroup of $\mathbf{\Gamma}$. If G contains no elements of finite period then it is free, by Theorem VIII.3. Suppose that G contains an element u of finite period. Then u is conjugate to x, y, or y^2, by Theorem VIII.4. Since G is normal, it must contain x, y, or y^2; and since $y^2 = y^{-1}$, it must contain x or y.

Suppose that G contains x. Then $G \supset \Delta(x)$, and so G can only be $\mathbf{\Gamma}$ or $\mathbf{\Gamma}^2$. Suppose that G contains y. Then G contains $\Delta(y)$, and so G can only be $\mathbf{\Gamma}$ or $\mathbf{\Gamma}^3$. This completes the proof.

11. The rank of a subgroup

We can now use the Schreier theorem to determine the rank of any free normal subgroup of $\mathbf{\Gamma}$ of finite index. Schreier's result states that if F is a free group of rank r, and H is a subgroup of F of index n, then H is a free group of rank $1 + n(r - 1)$. This theorem is not directly applicable to $\mathbf{\Gamma}$, since $\mathbf{\Gamma}$ is not free, but we can circumvent this by the following argument:

Suppose that G is a free normal subgroup of $\mathbf{\Gamma}$ of finite index μ. Then $\mathbf{\Gamma}' \cap G$ is also a free normal subgroup of $\mathbf{\Gamma}$ of finite index, and we have

$$\mathbf{\Gamma}' \supset \mathbf{\Gamma}' \cap G, \tag{9}$$

$$G \supset \mathbf{\Gamma}' \cap G \tag{10}$$

Put $(\mathbf{\Gamma}' : \mathbf{\Gamma}' \cap G) = m$, $(G : \mathbf{\Gamma}' \cap G) = n$. Let r be the rank of G, R the rank of $\mathbf{\Gamma}' \cap G$. Since the rank of $\mathbf{\Gamma}'$ is 2, the Schreier theorem applied to the pairs (9), (10) gives

$$R - 1 = m(2 - 1), \qquad R - 1 = n(r - 1)$$

so that

$$r = 1 + m/n$$

On the other hand,

$$\frac{m}{n} = \frac{(\mathbf{\Gamma}' : \mathbf{\Gamma} \cap G)}{(G : \mathbf{\Gamma}' \cap G)} = \frac{(\mathbf{\Gamma} : \mathbf{\Gamma}' \cap G)/(\mathbf{\Gamma} : \mathbf{\Gamma}')}{(\mathbf{\Gamma} : \mathbf{\Gamma}' \cap G)/(\mathbf{\Gamma} : G)} = \frac{\mu}{6}$$

since $(\mathbf{\Gamma} : \mathbf{\Gamma}') = 6$. Thus $r = 1 + \mu/6$, and we have proved

Theorem VIII.7. Let G be a normal subgroup of $\mathbf{\Gamma}$ of finite index μ, $G \neq \mathbf{\Gamma}, \mathbf{\Gamma}^2, \mathbf{\Gamma}^3$. The G is a free group of rank r, where

$$r = 1 + \mu/6 \tag{11}$$

We note that μ must be divisible by 6, which can also be seen directly, since xG, yG must occur as distinct cosets $\neq G$ in a coset decomposition of Γ modulo G. Notice that the above discussion actually implies that if G is *any* free subgroup of finite index μ, then its rank r is given by (11).

12. Definition of level of a normal subgroup

Another important number associated with the normal subgroup G is defined as follows: Let $z = xy$ be the element of Γ corresponding to S. Then the least positive integer n such that z^n belongs to G is said to be the *level* of G. Clearly, $n \mid \mu$ since z^μ belongs to G.

13. Classification by level

The groups $\Delta(z^n)$ are of considerable interest. The quotient groups $\Gamma/\Delta(z^n) = Q_n$ have defining relations

$$x^2 = y^3 = (xy)^n = 1$$

and are well known as a class of binary polyhedral groups (see [5]). The groups are infinite for $n \geq 6$. For $n \leq 5$, their orders are as follows:

(12)

n	1	2	3	4	5
$o(Q_n)$	1	6	12	24	60

Since any normal subgroup of finite index and level n must contain $\Delta(z^n)$, the enumeration of the normal subgroups of levels ≤ 5 becomes an easy matter. They are given in the following table:

n	normal subgroups of finite index and level n
1	Γ
2	Γ^2, $\Delta(z^2)$
3	Γ^3, $\Delta(z^3)$
4	$\Delta(z^4)$
5	$\Delta(z^5)$

There are infinitely many normal subgroups of level 6, which have all been described [45]. There is a one-to-one correspondence between such subgroups G and triples of integers p, m, d which satisfy

$$(13) \qquad p > 0, \qquad 0 \leq m \leq d - 1, \qquad m^2 + m + 1 \equiv 0 \bmod d$$

The group G consists of all words w of Γ which satisfy

$$(14) \qquad e(w; a) \equiv 0 \bmod p, \qquad e(w; b) \equiv me(w; a) \bmod dp$$

where a, b are the generators of Γ'.G always contains Γ'', the second commutator subgroup of Γ, and Γ'' may be characterized as $\Delta(z^6)$.

This scheme of classification may be continued, and further results of a similar nature may be found in [49].

14. Congruence groups

Much of the interest of the theory is centered on the congruence groups, which of course must be defined in terms of the matrix representation of Γ, and which have been discussed generally in Chapter VII for all dimensions. Contrary to the results of Mennicke mentioned in Chapter VII, there are infinitely many subgroups of Γ of finite index which are not congruence groups. We intend to discuss this question in some detail.

Because of the transition from $SL(2, Z)$ to $PSL(2, Z)$, it is more convenient to define the principal congruence subgroup of Γ of level n as the set of linear fractional transformations (1) such that

$$a \equiv d \equiv \pm 1 \bmod n, \qquad b \equiv c \equiv 0 \bmod n$$

As before, we denote this group by $\Gamma(n)$. Then again as before, a congruence group is any subgroup of Γ which contains some $\Gamma(n)$.

The results of Chapter VII imply that

$$(15) \qquad \mu(n) = (\Gamma : \Gamma(n)) = \begin{cases} \frac{1}{2} n^3 \prod_{p \mid n} \left(1 - \frac{1}{p^2}\right), & n > 2 \\ 6 & n = 2 \end{cases}$$

Notice that there is agreement in the two uses of the word level: Clearly, n is the least positive integer such that

$$\pm S^n = \pm \begin{bmatrix} 1 & n \\ 0 & 1 \end{bmatrix}$$

belongs to $\Gamma(n)$.

The first few values of $\mu(n)$ are as follows:

(16)

n	1	2	3	4	5
$\mu(n)$	1	6	12	24	60

Because of the agreement of the values of the tables (12) and (16), and because $\Gamma(n) \supset \Delta(S^n)$, we must have

$$\Gamma(n) = \Delta(S^n), \qquad 1 \leq n \leq 5$$

There is a useful theorem by Wohlfahrt (see [71]) which provides an effective criterion for deciding whether or not an arbitrary subgroup of finite index is a congruence group. We require some preliminary definitions before stating the theorem.

15. Parabolic elements

An element A of Γ is said to be *parabolic* if $tr(A) = \pm 2$. It is a simple exercise to show that any parabolic element must be of the form

$$A = \pm C^{-1} S^t C$$

where C belongs to Γ and t is an integer.

16. Definition of level of any subgroup

Now let G be any subgroup of Γ of index μ. Then if A is any element of Γ, $A^{\mu!}$ belongs to G (this is a consequence of a simple argument based on the remark that there can be no more than μ distinct powers of A modulo G). For a given element C of Γ, let $e = e(C)$ be the least positive integer such that $\pm C^{-1} S^e C \in G$. Then the possible values for e are finite in number, since $e \mid \mu!$. Let n be the least common multiple of all the integers $e(C)$, as C runs over Γ. Then we shall say that G is of *level* n.

Since $e(C)$ divides n for every $C \in \Gamma$, G must contain every conjugate of S^n. Thus $G \supset \Delta(S^n)$. Conversely, if n is the least positive integer such that $G \supset \Delta(S^n)$, then G is of level n.

This definition of level agrees with the previous ones when G is normal, since $e(C)$ is then independent of C. Then Wohlfahrt's criterion is that G is a congruence group if and only if G contains $\Gamma(n)$.

17. Wohlfahrt's theorem

We shall give a proof of Wohlfahrt's theorem which is purely algebraic, and in this respect is somewhat simpler than the original proof. We first prove a lemma.

Lemma VIII.1. Let G be a subgroup of $\mathbf{\Gamma}$ such that

$$G \supset \mathbf{\Gamma}(mn), \tag{17}$$

$$G \supset \mathbf{\Delta}(S^n) \tag{18}$$

where m, n are arbitrary positive integers. Then $G \supset \mathbf{\Gamma}(n)$.

Proof. We first note that (18) implies that

$$S(nx) = \begin{bmatrix} 1 & nx \\ 0 & 1 \end{bmatrix}, \qquad W(ny) = \begin{bmatrix} 1 & 1 \\ ny & 0 \end{bmatrix},$$

$$V(z) = \begin{bmatrix} 1 + nz & -n \\ nz^2 & 1 - nz \end{bmatrix}$$

are all in G for any integral values of x, y, z. Let

$$M = \pm \begin{bmatrix} 1 + na & * \\ nc & * \end{bmatrix}$$

be any element of $\mathbf{\Gamma}(n)$. Then x may be determined so that $(1 + na + n^2cx, m) = 1$, since $(1 + na, n^2c) = 1$. For this x,

$$M_1 = S(nx)M = \pm \begin{bmatrix} 1 + na_1 & * \\ nc & * \end{bmatrix}$$

where $(1 + na_1, m) = 1$. Now consider

$$M_2 = V(z)M_1$$

Then the (1, 1) element of M_2 is

$$\pm\{(1 + nz)(1 + na_1) - n^2c\}$$

We wish to determine z so that

$$(1 + nz)(1 + na_1) - n^2c \equiv 1 \bmod mn$$

which is equivalent to

$$a_1 - nc + (1 + na_1)z \equiv 0 \bmod m$$

Since $(1 + na_1, m) = 1$, this has a solution z. For this z,

$$M_2 \equiv \pm \begin{bmatrix} 1 & nb_2 \\ nc_2 & * \end{bmatrix} \bmod mn$$

Hence $M_2 \equiv \pm W(nc_2)S(nb_2) \bmod mn$, and so $S(-nb_2)W(-nc_2)M_2 \equiv \pm I \bmod mn$. It follows that $S(-nb_2)W(-nc_2)M_2$ belongs to G, because of (17). Backtracking, we get that M belongs to G. This completes the proof.

We now prove Wohlfahrt's theorem:

Theorem VIII.8 (Wohlfahrt). Let G be a subgroup of Γ of level n. Then G is a congruence group if and only if $G \supset \Gamma(n)$.

Proof. Certainly, if $G \supset \Gamma(n)$, then it is a congruence group. Conversely, suppose that G is a congruence group. Then there is a positive integer m such that $G \supset \Gamma(m)$, and hence $G \supset \Gamma(mn)$ as well. Since G is of level n, it also contains $\Delta(S^n)$. But now the preceding lemma implies that $G \supset \Gamma(n)$. This completes the proof.

18. Consequences of Wohlfahrt's theorem

Let us apply this result to the normal subgroups G of level 6 described by (13) and (14). Then G is a congruence group if and only if it contains $\Gamma(6)$. Since G is of index $6dp^2$, and since $\Gamma(6)$ is of index 72, the values of d and p are limited by the requirement $dp^2 | 12$. All possible cases when $G \supset \Gamma(6)$ are easily determined, and we find

Theorem VIII.9. The normal subgroup G of level 6 with parameters p, m, d is a congruence group if and only if

$$(p, m, d) = (1, 0, 1), (1, 1, 3), (2, 0, 1), (2,1, 3)$$

Let us take another example based on $\Gamma(2)$, which is of index 6. According to Theorem VIII.7, $\Gamma(2)$ is a free group of rank 2. It is readily shown that

$$A = \begin{bmatrix} 1 & 2 \\ 0 & 1 \end{bmatrix}, \qquad B = \begin{bmatrix} 1 & 0 \\ 2 & 1 \end{bmatrix}$$

are free generators of $\Gamma(2)$.

Let $\Gamma_n(2)$ be the subgroup of $\Gamma(2)$ consisting of all words W of $\Gamma(2)$

for which

$$(19) \qquad e(W; A) \equiv e(W; B) \equiv 0 \bmod n$$

Then because S, T are generators of Γ, and

$$SAS^{-1} = A, \qquad SBS^{-1} = -AB^{-1},$$
$$TAT^{-1} = B^{-1}, \qquad TBT^{-1} = A^{-1},$$

we see that condition (19) is preserved under conjugation, so that $\Gamma_n(2)$ is a normal subgroup of Γ. $\Gamma_n(2)$ is of index n^2 in $\Gamma(2)$, $\Gamma(2)/\Gamma_n(2)$ is abelian, and the level of $\Gamma_n(2)$ is clearly $2n$. Thus $\Gamma_n(2)$ is a congruence group if and only if it contains $\Gamma(2n)$. Thus a necessary condition is

$$6n^2 \mid \mu(2n)$$

In order to simplify the determination of the possible cases, we argue as follows: Suppose that $\Gamma_n(2)$ is a congruence group, and let p be any prime dividing n. Then $\Gamma_p(2)$ must also be a congruence group, since $\Gamma_p(2) \supset \Gamma_n(2)$. Thus p must satisfy

$$6p^2 \mid \mu(2p)$$

If p is odd, this becomes

$$2p \mid p^2 - 1$$

which is impossible. Thus we get

Theorem VIII.10. Suppose that n is not a power of 2. Then $\Gamma_n(2)$ is not a congruence group.

Many other examples of this type may be given: See papers [63], [47], and [51] for related material. All normal congruence groups were determined in [43] and [30].

It is worthwhile giving one fairly comprehensive class of examples. Choose G as any free subgroup of index μ and level n. Then G is of rank $r = 1 + \mu/6 \geq 2$. Let $g_1, g_2, \ldots, g_r$ be free generators of G. For any positive integer p, define G_p as the subgroup of Γ consisting of all words w such that

$$e(w; g_1) \equiv e(w; g_2) \equiv 0 \bmod p$$

Then G/G_p is abelian of order p^2, and if $u \in G$, then $u^p \in G_p$. It follows that G_p is of index μp^2, and of level n or np.

In order to show that G_p is not a congruence group, it will be sufficient to show that G_p does not contain $\Gamma(np)$; and in order to insure that this is so, it will be sufficient to choose p so that

$$\mu(np) \not\equiv 0 \bmod p^2 \tag{20}$$

If p is chosen prime and larger than $n + 1$, then $(\mu(n), p) = 1$ and (20) is satisfied, since $\mu(np) = \mu(n) \cdot p(p^2 - 1)$. Thus we have

Theorem VIII.11. Let G be any free subgroup of index μ and level n. Let $g_1, g_2, \ldots, g_r$ be free generators of G, where $r = 1 + \mu/6 \geq 2$. Let G_m be the subgroup of G consisting of all words w of G satisfying

$$e(w; g_1) \equiv e(w; g_2) \equiv 0 \bmod m$$

Then if m is divisible by a prime $> n + 1$, G_m is not a congruence group.

19. The parabolic class number

We now resume our study of the parabolic elements of the normal subgroup G of index μ and level n. Let P be the infinite cyclic subgroup $\{z^n\}$ (or $\{S^n\}$, depending on whether we are referring to the abstract group or its matrix representation). We consider the conjugates of P, $u^{-1}Pu$, as u runs over Γ. The number of distinct conjugates over G is known as the *parabolic class number* of G, and is denoted by t. We shall prove

Theorem VIII.12. The parabolic class number of the normal subgroup G of index μ and level n is

$$t = \mu/n$$

Proof. Let H be the group generated by z and G; $H = \{z, G\}$. Then $(H : G) = n$, G is a normal subgroup of H, and

$$H = G + zG + \cdots + z^{n-1}G \tag{21}$$

is a left coset decomposition for H modulo G.

Thus $(\Gamma : H) = \mu/n = t$, and so we have a right coset decomposition for Γ modulo H,

$$\Gamma = Hx_1 + Hx_2 + \cdots + Hx_t \tag{22}$$

Because of (21) and (22), we can show that

$$\Gamma = \sum_{\substack{0 \leq j \leq n-1 \\ 1 \leq i \leq t}} z^j x_i G \tag{23}$$

is a left coset decomposition for Γ modulo G. For there are the proper number of cosets; and if

$$(z^j x_i)^{-1}(z^k x_l) \in G,$$

then

$$x_i^{-1} z^{k-j} x_l \in G;$$

and since G is normal,

$$z^{k-j} x_l x_i^{-1} \in G$$

Thus

$$x_l x_i^{-1} \in H$$

which implies that $l = i$. It then follows that $z^{k-j} \in G$, so that $k = j$. Thus we have shown that (23) is indeed a left coset decomposition for Γ modulo G.

Now let u be any element of Γ. Then u may be written as

$$u = z^j x_i g, \qquad g \in G$$

Thus

$$u^{-1} P u = g^{-1} x_i^{-1} z^{-j} P z^j x_i g = g^{-1} x_i^{-1} P x_i g$$

so that $u^{-1} Pu$ is conjugate over G to one of the t groups

$$x_i^{-1} P x_i, \qquad 1 \leq i \leq t$$

But no two of these are conjugate over G. For if

$$x_i^{-1} P x_i = g^{-1} x_j^{-1} P x_j g, \qquad g \in G$$

then

$$(x_j g x_i^{-1}) P = P(x_j g x_i^{-1})$$

which implies that $x_j g x_i^{-1} = z^k$ (this is proved most quickly via the matrix representation). It follows that

$$x_j x_i^{-1} = x_j g^{-1} x_j^{-1} \cdot z^k$$

so that $x_j x_i^{-1}$ belongs to G, since $x_j g^{-1} x_j^{-1}$ belongs to G. Thus' $i = j$ and the proof is complete.

20. Related class numbers

Now that we have defined and determined the parabolic class number for a normal subgroup, we generalize this idea, together with certain important related ones, to any subgroup G of finite index μ. Let

$$\Gamma = x_1G + x_2G + \cdots + x_\mu G$$

be a left coset decomposition for Γ modulo G. Thus any element u of Γ may be written as

$$u = x_i g, \qquad g \in G$$

Let w be a fixed element of Γ other than 1(w will be chosen as x, y, or z later). Then the groups

$$u^{-1}\{w\}u, \qquad u \in G$$

fall into at most μ conjugacy classes over G; namely,

$$x_i^{-1}\{w\}x_i, \qquad 1 \leq i \leq \mu$$

Of course, not all of these are subgroups of G. Let n_i be the exponent of $x_i^{-1}wx_i$ modulo G. Then

$$x_i^{-1}\{w^{n_i}\}x_i \subset G, \qquad 1 \leq i \leq \mu \tag{24}$$

However, these are not necessarily distinct over G, or different from the trivial group $\{1\}$. Denote the number of groups (24) which are different from $\{1\}$ and which are distinct over G by $f(w)$. Then we make the following definitions:

$f(z) = t = t(G) =$ the number of *parabolic classes* of G;
$f(x) = e_2 = e_2(G) =$ the number of *elliptic classes* of period 2 of G;
$f(y) = e_3 = e_3(G) =$ the number of *elliptic classes* of period 3 of G.

Because $\{x\}$ and $\{y\}$ are cyclic of prime order, and so have no proper subgroups, the numbers e_2, e_3 are easily determined. We have

Theorem VIII.13. The elliptic class number e_2 is just the number of conjugates

$$x_i^{-1}xx_i, \qquad 1 \leq 1 \leq \mu$$

which belong to G; and e_3 similarly is just the number of conjugates

$$x_i^{-1}yx_i, \qquad 1 \leq 1 \leq \mu$$

which belong to G.

Proof. It will be sufficient to prove the first statement only, since the argument is the same in either case. Since $\{x\}$ is cyclic of prime order, the only possible values for the n_i of (24) are 1, if the group $x_i^{-1}\{x^{n_i}\}x_i$ is to be different from $\{1\}$. It is only necessary to show therefore that if $x_i^{-1}xx_i$, $x_j^{-1}xx_j$ belong to G, then they are conjugate over G if and only if $i = j$. Suppose then that for some $g \in G$,

$$x_i^{-1}xx_i = g^{-1}x_j^{-1}xx_jg$$

Then $x_jgx_i^{-1}$ commutes with x, which implies that

$$x_jgx_i^{-1} = x^k$$

Thus

$$x_i^{-1}x_j = g^{-1}(x_j^{-1}xx_j)^k$$

so that $x_i^{-1}x_j \in G$. Hence $i = j$, and the proof is complete.

The significance of the numbers t, e_2, e_3, may be somewhat obscure, but they have a natural geometric meaning, which will emerge in the discussion that follows.

21. The geometric approach

We now take the geometrical approach, and regard Γ as the group of linear fractional transformations (1) acting on the complex τ-plane. Since Γ is generated by

$$S\tau = \tau + 1, \qquad T\tau = -1/\tau$$

and since each of these maps the interior of the upper τ half-plane onto itself, the real axis onto itself, and the interior of the lower τ half-plane onto itself, the same is true of any element of Γ. We may therefore confine ourselves to the action of Γ on the upper τ half-plane, which we denote by $\mathcal{H}$.

We first introduce some terminology, which was partially described previously. Let A be any element of Γ. Then we say that

A is *elliptic*, if $|tr(A)| < 2$;
A is *parabolic*, if $|tr(A)| = 2$;
A is *hyperbolic*, if $|tr(A)| > 2$.

Then the elements of finite period of Γ coincide with the elliptic elements, so that any elliptic element is conjugate over Γ to $\pm T$,

$\pm U$, $\pm U^2$. The parabolic elements are conjugate over Γ to $\pm S^k$. The hyperbolic elements have no special properties of this kind.

Let us call two points τ_1, τ_2 of $\mathcal{H}$ *Γ-equivalent* if there is an element $A = \left[\begin{smallmatrix} a & b \\ c & d \end{smallmatrix}\right]$ of Γ such that

$$\tau_2 = A\tau_1 = \frac{a\tau_1 + b}{c\tau_1 + d}$$

Then this is an equivalence relationship, which partitions the points of $\mathcal{H}$ into equivalence classes. A set of points, one from each equivalence class, is called a *fundamental set*. We now describe the standard fundamental set associated with Γ.

We first note that every point of $\mathcal{H}$ is Γ-equivalent to one in the strip

$$-\tfrac{1}{2} \leq \mathrm{re}(\tau) \leq \tfrac{1}{2}$$

because the translation $S\tau = \tau + 1$ belongs to Γ. We next note that any point of $\mathcal{H}$ is also Γ-equivalent to one outside the unit circle; that is, to a point in

$$|\tau| \geq 1$$

because the transformation $T\tau = -1/\tau$ belongs to Γ. By repeated use of these two transformations, it is easy to show that any point of $\mathcal{H}$ is Γ-equivalent to one satisfying

$$-\tfrac{1}{2} \leq \mathrm{re}(\tau) \leq \tfrac{1}{2}, \qquad |\tau| \geq 1$$

The region defined by

$$(25) \qquad -\tfrac{1}{2} < \mathrm{re}(\tau) < \tfrac{1}{2}, \qquad |\tau| > 1$$

together with its boundary with Γ-equivalent points identified, may be shown to be a fundamental set. The set of points (25) is the standard *fundamental region* of Γ, which we denote by $\mathcal{R} = \mathcal{R}(\Gamma)$.

The boundary of $\mathcal{R}$ contains four distinguished points: the point $\tau = i\infty$, which is a fixed point of the parabolic transformation $S\tau = \tau + 1$; the point $\tau = i$, which is a fixed point of the elliptic transformation $T\tau = -1/\tau$ of period 2; and the Γ-equivalent points $\tau = \rho$, $\tau = \rho^2$, $\rho = ei\pi/3$, which are fixed points of the elliptic transformations $U\tau = -1/(\tau + 1)$, $U^2\tau = -(\tau + 1)/\tau$, of period 3.

Now let G be any subgroup of Γ of finite index μ. Then the same considerations apply to G, and a fundamental region $\mathcal{R}(G)$ may be constructed by the following method, due to Ford (see [11]):

Let $A = \pm\left[\begin{smallmatrix} a & b \\ c & d \end{smallmatrix}\right]$ be any element of $\mathbf{\Gamma}$ such that $c \neq 0$. We define the *isometric* circle of A as the circle

$$|c\tau + d| \leq 1$$

Let $\mathfrak{S} = \mathfrak{S}(G)$ be that part of the upper τ half-plane which is exterior to every isometric circle of A, as A runs over the elements of G for which $c \neq 0$.

Let n be the exponent of $\mathfrak{S}$ modulo G. Then the intersection of $\mathfrak{S}$ with the strip

$$-\frac{n}{2} < \operatorname{re}(\tau) < \frac{n}{2}$$

is the fundamental region $\mathfrak{R}(G)$. Once again, a fundamental set is obtained by including the boundary with G-equivalent points identified.

The numbers t, e_2, e_3 defined previously now have the following meaning: t is the number of G-inequivalent points on the boundary of $\mathfrak{R}(G)$ which are fixed points of parabolic transformations of G, e_2 is the number of G-inequivalent points on the boundary of $\mathfrak{R}(G)$ which are fixed points of elliptic transformations of period 2 of G, and e_3 is the number of G-inequivalent points on the boundary of $\mathfrak{R}(G)$ which are fixed points of elliptic transformations of period 3 of G.

22. Genus

When corresponding sides of $\mathfrak{R}(G)$ are identified, $\mathfrak{R}(G)$ becomes a surface, whose genus g may be computed by the Euler–Poincaré formula. It turns out that

$$g = 1 + \frac{\mu}{12} - \frac{t}{2} - \frac{e_2}{4} - \frac{e_3}{3} \tag{26}$$

If G is free, for example, (26) becomes

$$g = 1 + \frac{\mu}{12} - \frac{t}{2}$$

Thus if G is a normal subgroup $\neq \mathbf{\Gamma}, \mathbf{\Gamma}^2, \mathbf{\Gamma}^3$ of index μ and level n, the genus becomes

$$g = 1 + \frac{\mu}{12} - \frac{\mu}{2n} = 1 + \frac{\mu(n-6)}{12n} \tag{27}$$

The normal subgroups of level 6 discussed previously are thus just those of genus 1. The case $g = 1$ is truly exceptional. It is the only value of g such that there are infinitely many normal subgroups of finite index of that genus. For any other value, the possible values for μ are limited, and so only finitely many instances can occur, since a finitely generated group possesses only finitely many subgroups of a given finite index. For example, the only normal subgroups of genus 0 are

$$\Gamma, \Gamma^2, \Gamma^3, \Gamma(2), \Gamma(3), \Gamma(4), \Gamma(5)$$

A classification of the normal subgroups by index and genus is given in [49].

23. Canonical generators and relations

On the basis of formula (26) and the fundamental region $\mathcal{R}(G)$, a canonical set of generators and relations for G may be obtained. It may be shown that G is generated by

e_2 elliptic elements of period 2, $E_1, E_2, \ldots, E_{e_2}$;
e_3 elliptic elements of period 3, $F_1, F_2, \ldots, F_{e_3}$;
t parabolic elements, $P_1, P_2, \ldots, P_t$;
$2g$ hyperbolic elements, $A_1, B_1, A_2, B_2, \ldots, A_g, B_g$;

with defining relations

$$E_1{}^2 = E_2{}^2 = \cdots = E_{e_2}^2 = F_1{}^3 = F_2{}^3 = \cdots = F_{e_3}^3 = 1$$

$$\prod_{i=1}^{e_2} E_i \prod_{j=1}^{e_3} F_j \prod_{k=1}^{t} P_k \prod_{l=1}^{g} \{A_l B_l A_l^{-1} B_l^{-1}\} = 1$$

The rank of G is thus

$$r = e_2 + e_3 + t + 2g - 1 = 1 + \tfrac{1}{6}\mu + \tfrac{1}{2}e_2 + \tfrac{1}{3}e_3$$

24. Related groups

We leave the modular group now and go on to some related groups. Our starting point is the observation that $PSL(2, Z)$ really provides a convenient representation for the abstract free product of a cyclic group of order 2 and a cyclic group of order 3. In this connection it is natural to look for similar representations of the free product of any pair of cyclic groups. The discussion that follows has this object.

25. Discrete groups

We will consider subgroups of $SL(2, R)$, where R is the real field. The really interesting subgroups are the discrete ones, and we will limit ourselves to the construction of such subgroups. A subgroup of $SL(2, R)$ is said to be *discrete* if it contains no convergent infinite sequence of distinct matrices. We will be concerned with groups generated by two elements. If A, B are arbitrary elements of $SL(2, R)$, then

$$G = \{A, B\}$$

denotes the subgroup of $SL(2, R)$ generated by A and B.

26. A class of free, discrete groups

We first look for criteria to ensure that G is a free, discrete group. Each element of G has the form

$$W = A^{r_1}B^{s_1}A^{r_2}B^{s_2} \cdots A^{r_n}B^{s_n} \tag{28}$$

where the exponents are different from 0 except possibly for r_1 and s_n. If all the exponents are different from 0, we say that W is of *type* (AB). A simple argument shows that

(a) G is free and freely generated by A and B if and only if A and B are not of finite period and no word of type (AB) with $n > 0$ represents the identity;

(b) G is discrete if and only if it contains no convergent infinite sequence $W_1, W_2, \ldots$ of distinct words W_i of type (AB).

The method consists of deriving inequalities fur the elements of the matrices A^rB^s, $rs \neq 0$. The inequalities carry over on multiplication and imply the desired results.

The theorem we intend to prove is as follows:

Theorem VIII.14. Let A, B be elements of $SL(2, R)$. Suppose that

$$A = \begin{bmatrix} -a & b \\ -c & d \end{bmatrix}, \qquad B = \begin{bmatrix} -\alpha & -\beta \\ \gamma & \delta \end{bmatrix}$$

where $a, b, c, d, \alpha, \beta, \gamma, \delta \geq 0$, and $t = d - a \geq 2$, $\tau = \delta - \alpha \geq 2$. Then the group $G = \{A, B\}$ is a free discrete subgroup of $SL(2, R)$ and is freely generated by A, B.

We first prove two lemmas.

Lemma VIII.2. There is a conjugacy over $GL(2, R)$ that takes A into $\left[\begin{smallmatrix}0 & 1\\ -1 & t\end{smallmatrix}\right]$ and B into a matrix with the same sign pattern as B.

Proof. Since $bc = 1 + ad \geq 1$, c cannot vanish. Put

$$M = \begin{bmatrix} c & -a \\ 0 & 1 \end{bmatrix} \in GL(2, R)$$

Then

$$MAM^{-1} = \begin{bmatrix} 0 & 1 \\ -1 & t \end{bmatrix},$$

$$MBM^{-1} = \frac{1}{c}\begin{bmatrix} -\alpha c - \gamma a & -\beta c^2 - \gamma a^2 - (\alpha + \delta)ac \\ \gamma & \gamma a + \delta c \end{bmatrix}$$

Since MBM^{-1} and B obviously have the same sign pattern, the lemma follows.

Because of this lemma, we lose no generality in assuming that $A = \left[\begin{smallmatrix}0 & 1\\ -1 & t\end{smallmatrix}\right]$, and this will be done in what follows. As usual, we define

$$\operatorname{sgn}(x) = \begin{cases} 1, & x \geq 0 \\ -1, & x < 0 \end{cases}$$

Let X, Y be any elements of $SL(2, R)$. We write $X \gg Y$ to mean that every element of X is nonnegative and exceeds or equals the absolute value of the corresponding element of Y. Notice that if $X_1 \gg Y_1$ and $X_2 \gg Y_2$, then $X_1 X_2 \gg Y_1 Y_2$.

We now prove

Lemma VIII.3. Let r and s be nonzero integers. Then

$$\operatorname{sgn}(rs)A^r B^s \gg |rs| \begin{bmatrix} \gamma & 0 \\ 0 & \beta \end{bmatrix} \tag{29}$$

Furthermore, if $\alpha = 0$, $\beta = 1$, $\gamma = 1$ (so that $B = \left[\begin{smallmatrix}0 & -1\\ 1 & \tau\end{smallmatrix}\right]$), then

$$\operatorname{sgn}(rs)A^r B^s \gg |rs|\, C(r, s) \tag{30}$$

where $C(r, s)$ is one of the matrices

$$\begin{bmatrix} 1 & 2 \\ 0 & 1 \end{bmatrix}, \quad \begin{bmatrix} 1 & 0 \\ 2 & 1 \end{bmatrix}, \quad \begin{bmatrix} 2 & 1 \\ 1 & 1 \end{bmatrix}, \quad \begin{bmatrix} 1 & 1 \\ 1 & 2 \end{bmatrix}$$

Proof. Since A has trace t and determinant 1, A satisfies its characteristic equation

$$A^2 = tA - I$$

Hence for each integer r, A^r must be a linear combination of A and I. In fact,

$$A^r = t_r A - t_{r-1} I$$

where $t_0 = 0$, $t_1 = 1$, $t_{r+1} = tt_r - t_{r-1}$. It is readily seen that

(31) $$t_{-r} = -t_r$$

and (because $t \geq 2$)

(32) $$t_r \geq r, \qquad r \geq 0$$

Similarly for each integer s,

$$B^s = \tau_s B - \tau_{s-1} I$$

where $\tau_0 = 0$, $\tau_1 = 1$, $\tau_{s+1} = \tau\tau_s - \tau_{s-1}$, and where

(33) $$\tau_{-s} = -\tau_s,$$

(34) $$\tau_s \geq s, \qquad s \geq 0$$

Direct computation now shows that

$$A^r B^s = \begin{bmatrix} a_{r,s} & b_{r,s} \\ c_{r,s} & d_{r,s} \end{bmatrix}$$

where

$$\begin{aligned} a_{r,s} &= t_{r-1}\tau_{s-1} + \alpha t_{r-1}\tau_s + \gamma t_r \tau_s, \\ b_{r,s} &= \beta t_{r-1}\tau_s + \alpha t_r \tau_s + t_r \tau_{s+1}, \\ c_{r,s} &= t_r \tau_{s-1} + \alpha t_r \tau_s + \gamma t_{r+1}\tau_s, \\ d_{r,s} &= \beta t_r \tau_s + \alpha t_{r+1}\tau_s + t_{r+1}\tau_{s+1} \end{aligned}$$

Observe that if $rs > 0$, then all the elements of $A^r B^s$ are nonnegative, and that if $rs < 0$, then all the elements of $-A^r B^s$ are nonnegative. In fact, using (31), (32), (33), and (34), we obtain

$$A^r B^s \gg \quad rs \begin{bmatrix} \gamma & \alpha + 1 \\ \alpha + \gamma & \alpha + \beta + 1 \end{bmatrix}, \qquad r > 0, \quad s > 0,$$

$$-A^r B^s \gg -rs \begin{bmatrix} \gamma & \alpha \\ \alpha + \gamma + 1 & \alpha + \beta \end{bmatrix}, \qquad r > 0, \quad s < 0,$$

$$-A^rB^s \gg -rs\begin{bmatrix} \alpha+\gamma & \alpha+\beta+1 \\ \alpha & \beta \end{bmatrix}, \qquad r<0, \quad s>0,$$

$$A^rB^s \gg \quad rs\begin{bmatrix} \alpha+\gamma+1 & \alpha+\beta \\ \alpha+1 & \beta \end{bmatrix}, \qquad r<0, \quad s<0$$

from which both (29) and (30) follow. This completes the proof of the lemma.

We now turn to the proof of the theorem. If W is of type (AB), define $h(W)$ as the absolute value of the product of the exponents of W. We note first that neither A nor B is of finite period. Let W, given by (28), be any word of type (AB), and put

$$\varepsilon = \operatorname{sgn}(r_1 s_1) \operatorname{sgn}(r_2 s_2) \cdots \operatorname{sgn}(r_n s_n)$$

Then Lemma VIII.3 implies that

$$\varepsilon W \gg h(W)\begin{bmatrix} \gamma^n & 0 \\ 0 & \beta^n \end{bmatrix} \tag{35}$$

and also that if $B = [\begin{smallmatrix} 0 & -1 \\ 1 & \tau \end{smallmatrix}]$, then

$$\varepsilon W \gg h(W) C(r_1, s_1) C(r_2, s_2) \cdots C(r_n, s_n) \tag{36}$$

If $B \neq [\begin{smallmatrix} 0 & -1 \\ 1 & \tau \end{smallmatrix}]$ then $\beta > 1$ or $\gamma > 1$, and (35) implies that W can never be the identity. If $B = [\begin{smallmatrix} 0 & -1 \\ 1 & \tau \end{smallmatrix}]$, then (36) implies that W can never be the identity. Hence we have proved the first part of the theorem; namely, the group G is free.

Now suppose that

$$W_i = A^{r_{i1}} B^{s_{i1}} A^{r_{i2}} B^{s_{i2}} \cdots A^{r_{ik_i}} B^{s_{ik_i}}, \qquad i = 1, 2, 3, \ldots$$

is any infinite sequence of distinct elements of type (AB). Then certainly

$$\sum_{j=1}^{k_i} \{|r_{ij}| + |s_{ij}|\} \longrightarrow \infty \quad \text{as} \quad i \longrightarrow \infty \tag{37}$$

which implies that either $h(W_i) \to \infty$ as $i \to \infty$, or $k_i \to \infty$ as $i \to \infty$, or both. Put

$$\varepsilon_i = \operatorname{sgn}(r_{i1} s_{i1}) \operatorname{sgn}(r_{i2} s_{i2}) \cdots \operatorname{sgn}(r_{ik_i} s_{ik_i})$$

We see again that

$$\varepsilon_i W_i \gg h(W_i)\begin{bmatrix} \gamma^{k_i} & 0 \\ 0 & \beta^{k_i} \end{bmatrix} \tag{38}$$

and also that if $B = \left[\begin{smallmatrix} 0 & -1 \\ 1 & \tau \end{smallmatrix}\right]$, then

$$\mathcal{E}_i W_i \gg h(W_i) C(r_{i1}, s_{i1}) C(r_{i2}, s_{i2}) \cdots C(r_{ik_i}, s_{ik_i}) \tag{39}$$

A little reflection shows that because of (37), (38), and (39), some of the elements of $\mathcal{E}_i W_i$ become arbitrarily large as $i \longrightarrow \infty$. Hence no such sequence converges, and we have proved the second part of the theorem; namely, the group G is discrete. This completes the proof.

27. A theorem of free products

Essentially the same method allows us to prove the following generalization (we omit the proof):

Theorem VIII.15. Let A, B be elements of $PSL(2, R)$, and let p, q be integers ≥ 2. Suppose that

$$A = \begin{bmatrix} -a & b \\ -c & d \end{bmatrix}, \qquad B = \begin{bmatrix} -\alpha & -\beta \\ \gamma & \delta \end{bmatrix}$$

where $a, b, c, d, \alpha, \beta, \gamma, \delta \geq 0$. Put $t = d - a$, $\tau = \delta - \alpha$. Then the group $G = \{A, B\}$ generated by A and B is discrete and equal to $\{A\} * \{B\}$ (the free product of the indicated cyclic groups) in each of the following four cases:

$$\begin{array}{ll} t \geq 2, & \tau \geq 2, \\ t = 2\cos \pi/p, & \tau \geq 2, \\ t \geq 2, & \tau = 2\cos \pi/q, \\ t = 2\cos \pi/p, & \tau = 2\cos \pi/q \end{array}$$

28. A theorem on representations of Γ

On the basis of these theorems, it was found possible to classify all real discrete representations of the modular group, or more generally the free product of any two cyclic groups of finite order, by elements of $PSL(2, R)$; and this was done in [26] and [27]. We quote the principal result of [26]:

Theorem VIII.16. For each $\rho \geq 1$ set

$$R_\rho = \begin{bmatrix} 0 & -\rho \\ \dfrac{1}{\rho} & 1 \end{bmatrix}$$

Put $T = \left[\begin{smallmatrix} 0 & 1 \\ -1 & 0 \end{smallmatrix}\right]$, $G_\rho = \{T, R_\rho\}$. Then

(a) $G_\rho = \{T\} * \{R_\rho\}$, the free product of the cyclic group $\{T\}$ of order 2 and the cyclic group $\{R_\rho\}$ of order 3.

(b) G_ρ is a discrete group.

(c) Every discrete faithful representation of Γ by a subgroup of $PSL(2, R)$ is conjugate over $GL(2, R)$ to G_ρ for some $\rho \geq 1$.

(d) If $\rho, \sigma \geq 1$, then G_ρ and G_σ are conjugate over $GL(2, R)$ if and only if $\rho = \sigma$.

EXERCISES AND PROBLEMS

1. Prove that the group $G = \{x, y\}$ with defining relations

$$x^2 = y^3 = (xy)^n = 1$$

is infinite for $n \geq 6$, by considering the homomorphism φ of G into $PSL(2, Q(\rho))$, $\rho = e^{i\pi/n}$, defined by

$$\varphi(x) = \pm \begin{bmatrix} 0 & 1 \\ -1 & 0 \end{bmatrix}, \qquad \varphi(y) = \pm \begin{bmatrix} 0 & -\rho \\ 1/\rho & 1 \end{bmatrix}$$

Hint: First verify that $\varphi(x)$, $\varphi(y)$, $\varphi(xy)$ are of periods 2, 3, n, respectively. Then show that $\varphi(xyx^{-1}y^{-1})$ is of trace $t = \pm(1 + 2\cos 2\pi/n)$, so that $|t| \geq 2$ for $n \geq 6$. Conclude that for $n \geq 6$, $\varphi(xyx^{-1}y^{-1})$ cannot be of finite period, whence $\varphi(G)$, and so G, must be infinite.

2. Suppose that α is transcendental. Prove that the group generated by

$$\begin{bmatrix} 1 & \alpha \\ 0 & 1 \end{bmatrix}, \qquad \begin{bmatrix} 1 & 0 \\ 1 & 1 \end{bmatrix}$$

is a free group.

Chapter IX

Finite Matrix Groups

The entire theory of finite groups is subsumed under this heading. Our emphasis will be on the matrix theory rather than the group theory. The most important theorems of this chapter come from the theory of group representations. Although it is not possible to give a complete presentation of this subject here, we will develop some of the key theorems as simply as possible in matrix terms. A detailed treatment along modern lines can be found in the book by Curtis and Reiner [6]. A simple development from the matrix standpoint can be found in Newman [52].

1. Reducibility and irreducibility

Let F be a field. Let $S = \{A\}$ be any set of matrices of F_n. Then S is said to be *reducible* if fixed positive integers r, s and a fixed matrix U of $GL(n, F)$ exist such that for each $A \in S$,

$$(1) \qquad U^{-1}AU = \begin{bmatrix} A_{11} & 0 \\ A_{21} & A_{22} \end{bmatrix}$$

where A_{11} is $r \times r$, A_{21} is $s \times r$, and A_{22} is $s \times s$. Otherwise S is said to be *irreducible*. Thus if V is any matrix of $GL(n, F)$, then $V^{-1}SV = \{V^{-1}AV\}$ is reducible if and only if S is reducible. For brevity, we say that S and $V^{-1}SV$ are *equivalent*.

2. Schur's lemma

The basic result on irreducible sets is due to Schur, and states the following:

Theorem IX.1 (Schur's lemma). Let $S = \{A\}$, $T = \{B\}$ be irreducible sets of matrices of F_n, F_m, respectively. Let M be a fixed $m \times n$ matrix over F which determines a one-to-one correspondence between S and T such that

$$MA = BM$$

Then either $M = 0$, or else $m = n$ and M is nonsingular.

Proof. Suppose that the rank of M is r, and write

$$M = P\begin{bmatrix} 0 & 0 \\ I_r & 0 \end{bmatrix}Q$$

where $P \in GL(m, F)$, $Q \in GL(n, F)$. Then for each $A \in S$ and the corresponding $B \in T$,

$$\begin{bmatrix} 0 & 0 \\ I_r & 0 \end{bmatrix}QAQ^{-1} = P^{-1}BP\begin{bmatrix} 0 & 0 \\ I_r & 0 \end{bmatrix} \tag{2}$$

Put

$$QAQ^{-1} = \begin{bmatrix} A_{11} & A_{12} \\ A_{21} & A_{22} \end{bmatrix}, \qquad P^{-1}BP = \begin{bmatrix} B_{11} & B_{12} \\ B_{21} & B_{22} \end{bmatrix}$$

where the partitioning is that imposed by (2). It is readily verified that A_{12} is $r \times (n - r)$ and B_{12} is $(m - r) \times r$. Then (2) implies that

$$A_{12} = 0, \qquad B_{12} = 0,$$

an impossibility, since S and T are irreducible. Since this can happen only vacuously, we must have that either $r = 0$ (in which case $M = 0$) or else $r = m = n$ (in which case M is nonsingular). This completes the proof.

An important corollary is the following:

Corollary IX.1. If F is algebraically closed, and M commutes with each matrix of the irreducible set S, then M is scalar.

Proof. Let λ be any eigenvalue of M. Then $M - \lambda I \in F_n$, is singular, and also commutes with each matrix of S. Thus $M - \lambda I$ must be 0, by Schur's lemma.

3. Burnside's theorem on irreducible sets of matrices

The most important single fact about irreducibility is the following theorem:

Theorem IX.2 (Burnside). Let F be algebraically closed. Let $G = \{A\}$ be a subgroup of $GL(n, F)$ which is irreducible as a set of matrices. Then any relationship

$$\sum_{p,q} \tau_{pq} a_{pq} = 0, \qquad \tau_{pq} \in F$$

can hold for all $A = (a_{pq})$ of G if and only if $\tau_{pq} = 0$ for all p, q.

Proof. Suppose that n^2 elements k_{qp} of F exist which are not all 0 such that

$$\sum_{p,q} k_{qp} a_{pq} = 0, \qquad \text{all } A \in G \tag{3}$$

Set $K = (k_{pq})$. Then (3) is equivalent to

$$\operatorname{tr}(KA) = 0, \qquad \text{all } A \in G \tag{4}$$

The totality of matrices K of F_n satisfying (4) is a vector space W over F of dimension $s \leq n^2$. Let $K_1, K_2, \ldots, K_s$ be a basis for W. Then every matrix K satisfying (4) is representable as a linear combination of $K_1, K_2, \ldots, K_s$. Furthermore if $K \in W$ then, for each $A \in G$, $KA \in W$, since

$$\operatorname{tr}(KA \cdot B) = \operatorname{tr}(K \cdot AB) = 0, \qquad \text{all } B \in G$$

Hence there are s elements $r_1, r_2, \ldots, r_s$ of F (which depend on A) such that

$$KA = \sum_{p=1}^{s} r_p K_p$$

and these are uniquely determined, since the matrices $K_1, K_2, \ldots, K_s$ are a basis. In particular we have

(5) $$K_pA = \sum_{q=1}^{s} r_{pq}K_q, \qquad 1 \leq p \leq s$$

Put

$$K_p = (k_{ij}^p), \qquad 1 \leq p \leq s$$

Then the s equations (5) become

(6) $$\sum_{l=1}^{n} k_{il}^p a_{lj} = \sum_{q=1}^{s} r_{pq}k_{ij}^q, \qquad 1 \leq p \leq s, \qquad 1 \leq i, j \leq n$$

Denote the $s \times s$ matrix (r_{ij}) by R, and the $s \times n$ matrix

$$\begin{bmatrix} k_{i1}^1 & k_{i2}^1 & \cdots & k_{in}^1 \\ k_{i1}^2 & k_{i2}^2 & \cdots & k_{in}^2 \\ \cdot & \cdot & & \cdot \\ \cdot & \cdot & & \cdot \\ \cdot & \cdot & & \cdot \\ k_{i1}^s & k_{i2}^s & \cdots & k_{in}^s \end{bmatrix}$$

by P_i, $1 \leq i \leq n$. Then the equations (6) may be written in the form

(7) $$P_iA = RP_i, \qquad 1 \leq i \leq n$$

Now suppose that $V = (v_{pq})$ is an arbitrary element of $GL(s, F)$. If we define

$$K_p' = \sum_{q=1}^{s} v_{pq}K_q, \qquad 1 \leq p \leq s$$

then $K_1', K_2', \ldots, K_s'$ is also a basis for W. Define R' by

$$R' = VRV^{-1} = (r'_{pq})$$

Then (5) takes the form

$$K_p'A = \sum_{q=1}^{s} r'_{pq}K_q', \qquad 1 \leq p \leq s$$

It follows that in our discussion the set $\{R\}$ may be replaced by any equivalent set, by making the appropriate change of basis in W.

We now appeal to Schur's lemma. Since the matrices $K_1, K_2, \ldots, K_s$ are a basis, none of them is 0. Hence not all of the matrices P_i can be 0. It follows that either P_i is a square nonsingular matrix for some i, in which case the sets $\{A\}$ and $\{R\}$ are equivalent, or the set $\{R\}$ is reducible.

Suppose first that P_i is square and nonsingular for some i so that $s = n$ and $\{A\}$ and $\{R\}$ are equivalent. Then after a suitable change of basis in W, we can in fact assume that they are the same. Equations (7) now become

$$P_i A = AP_i, \qquad 1 \leq i \leq n$$

Corollary IX.1 now implies that P_i is scalar:

$$P_i = k_i I, \qquad 1 \leq i \leq n$$

Thus

$$k_{ij}^p = k_i \delta_{pj}$$

Now from the fact that

$$\sum_{i,j} k_{ji}^p a_{ij} = 0, \qquad 1 \leq p \leq n$$

it follows that

$$\sum_{i,j} k_j \delta_{pi} a_{ij} = 0, \qquad 1 \leq p \leq n,$$

$$\sum_{j=1}^{n} k_j a_{pj} = 0, \qquad 1 \leq p \leq n$$

Hence

$$A \begin{bmatrix} k_1 \\ k_2 \\ \cdot \\ \cdot \\ \cdot \\ k_n \end{bmatrix} = 0$$

But A is nonsingular: Thus

$$k_j = 0, \qquad 1 \leq j \leq n$$

which in turn implies that all the matrices P_i are zero, $1 \leq i \leq n$; a contradiction.

Next, suppose that $\{R\}$ is reducible. Then after a suitable change of basis has been performed we may assume that for some positive integer t, every R has the form

$$R = \begin{bmatrix} S & 0 \\ T & U \end{bmatrix}$$

where the $t \times t$ matrices S are either all 0, or $\{S\}$ is irreducible. If $S = (s_{pq})$, then the first t equations of (5) becomes

$$K_p A = \sum_{q=1}^{t} s_{pq} K_q, \qquad 1 \leq p \leq t$$

If $\{S\}$ is irreducible, we conclude as before that $K_1, K_2, \ldots, K_t$ are all 0. If all the S are 0, the same conclusion can be drawn directly from the equations

$$K_pA = 0, \qquad 1 \leq p \leq t$$

since A is nonsingular. In either case we have arrived at a contradiction since no one of the matrices $K_1, K_2, \ldots, K_t$ can vanish. This concludes the proof.

An equivalent formulation is as follows:

Theorem IX.3. Let F be algebraically closed. Let G be an irreducible subgroup of $GL(n, F)$. Then the dimension of the vector space spanned by the matrices of G is n^2.

4. Burnside's theorem on groups of finite exponent

As an application of Burnside's theorem, we prove Burnside's celebrated result on groups of finite exponent. A group G is said to be of *exponent r* if $x^r = 1$ for every element x of G. Burnside's theorem follows:

Theorem IX.4 (Burnside). Let G be a subgroup of $GL(n, C)$ (C the complex field) of exponent r. Then G is finite.

Proof. Suppose first that G is irreducible. If $A \in G$, then $\mathrm{tr}(A)$ is the sum of n rth roots of unity, and so assumes only finitely many values (certainly not more than r^n). Therefore the set of traces

$$\{\mathrm{tr}(A) \colon A \in G\}$$

contains only finitely many distinct elements.

Since G is irreducible, there are elements $A_1, A_2, \ldots, A_{n^2}$ of G which are linearly independent. Let A be any element of G, and put $A = (a_{ij})$, $A_k = (a_{ij}^k)$, $1 \leq k \leq n^2$. Then

$$\mathrm{tr}(A_kA) = \sum_{i,j} a_{ij}^k a_{ji}, \qquad 1 \leq k \leq n^2$$

Regarding these as n^2 linear equations in the n^2 unknowns a_{ji}, we see that the rows of the coefficient matrix are linearly independent, and hence a unique solution exists, depending only on the coefficient matrix and the values $\operatorname{tr}(A_k A)$, $1 \leq k \leq n^2$. Since there are only finitely many possibilities for these, there are only finitely many possibilities for the a_{ji}. Thus the group is finite in this case.

We now proceed by induction on n. The theorem is certainly true for $n = 1$. Suppose the theorem proved for all dimensions $< n$. If G is irreducible, the discussion above shows that the theorem is true. If G is reducible, we may assume that G is already in reduced form; that is, that there are positive integers p, q such that every $A \in G$ has the form

$$A = \begin{bmatrix} B & 0 \\ C & D \end{bmatrix}$$

where B is $p \times p$, C is $q \times p$, and D is $q \times q$. Then the groups $\{B\}$ and $\{D\}$ are finite, by the induction hypothesis, since they are certainly of exponent r. Define

$$G_1 = \{A \in G \colon B = I\}, \qquad G_2 = \{A \in G \colon D = I\}$$

Then G_1, G_2 are normal subgroups of G of finite index. The same is therefore true of $G_1 \cap G_2$. Let $A \in G_1 \cap G_2$. Then

$$A = \begin{bmatrix} I & 0 \\ C & I \end{bmatrix}$$

Since $A^r = \left[\begin{smallmatrix} I & 0 \\ rC & I \end{smallmatrix}\right] = I$, $C = 0$. Hence $G_1 \cap G_2 = \{I\}$, and it follows that G is finite. This completes the proof.

Almost the same proof shows that if G is any subgroup of $GL(n, C)$ with only finitely many conjugacy classes, then G is finite. The relevant point is that the set of traces contains only finitely many distinct elements.

The theorem is false for arbitrary groups; for example, the direct product of countably many cyclic groups of order r, By a more complicated argument, Schur has shown that a finitely generated subgroup of $GL(n, C)$, in which each element is of finite period, is finite.

5. Matrix representations and characters

We now wish to derive some of the character relations for finite groups. For this purpose we change our notation slightly. Let $G = \{x\}$ be some fixed, finite group. Let $\alpha = \{A(x) \colon x \in G\}$ be a *matrix repre-*

sentation of G, where the matrices $A(x)$ belong to $GL(n, C)$, and satisfy

$$A(xy) = A(x)A(y), \qquad \text{all } x, y \in G$$

There is no need to assume that the representation is faithful: α may only be a homomorphic image of G. α is said to be of *degree n*. We set

$$\chi_\alpha(x) = \operatorname{tr}(A(x)), \qquad x \in G$$

The set

$$\{\chi_\alpha(x) : x \in G\}$$

is known as the *character* of the representation α.

6. Orthogonality relationships of the first kind

Suppose now that

$$\alpha = \{A(x) : x \in G\}, \qquad \beta = \{B(x) : x \in G\}$$

are two irreducible matrix representations of G. Let α be of degree n, β of degree m, and put $g = o(G)$. Let U be an arbitrary $n \times m$ complex matrix, and form

$$V = \sum_{x \in G} A(x)UB(x^{-1}) = \sum_{x \in G} A(x)UB(x)^{-1}$$

Let y be any element of G. Then

$$A(y)VB(y)^{-1} = \sum_{x \in G} A(yx)UB(yx)^{-1} = V$$

since yx runs over G as x does. Comparing elements in the (i, j) position, we find that

$$v_{ij} = \sum_{x \in G} a_{ip}(x)u_{pq}b_{qj}(x^{-1})$$

Suppose first that α and β are not equivalent. Then Schur's lemma implies that $V = 0$; and since U was arbitrary, we find that

(8) $$\sum_{x \in G} a_{ip}(x)b_{qj}(x^{-1}) = 0, \qquad \alpha, \beta \text{ not equivalent}$$

In particular, setting $p = i$, $q = j$ and summing over i, j we obtain

(9) $$\sum \chi_\alpha(x)\chi_\beta(x^{-1}) = 0, \qquad \alpha, \beta \text{ not equivalent}$$

Now consider the case when $\alpha = \beta$. Then V commutes with every element of α, and the corollary to Schur's lemma implies that V is scalar: $V = \lambda I$. To determine λ we compute the trace:

$$\mathrm{tr}(V) = \mathrm{tr}(\sum_{x\in G} A(x)\, U\, A(x)^{-1}) = \sum_{x\in G} \mathrm{tr}(A(x)\, U\, A(x)^{-1}) = g\,\mathrm{tr}(U)$$

Hence

$$\lambda = (g/n)\,\mathrm{tr}(U)$$

and so

$$(g/n)(u_{11} + u_{22} + \cdots + u_{nn})\,\delta_{ij} = \sum_{x\in G}\sum_{p,q} a_{ip}(x) u_{pq} a_{qj}(x^{-1})$$

Comparing coefficients, we find that

$$\sum_{x\in G} a_{ip}(x) a_{qj}(x^{-1}) = (g/n)\,\delta_{ij}\,\delta_{pq} \tag{10}$$

Setting $p = i$, $q = j$ and summing as before over all i, j we find that

$$\sum_{x\in G} \chi_\alpha(x)\chi_\alpha(x^{-1}) = g \tag{11}$$

The two formulas (9) and (11) are known as the orthogonality relationships of the first kind.

Formula (11) may be generalized slightly. Let y be any element of G. Multiply (10) by $a_{pl}(y)$ and sum over all p. The result, after some simplification, is

$$\sum_{x\in G} a_{qj}(x^{-1}) a_{il}(xy) = (g/n)\,\delta_{ij} a_{ql}(y) \tag{12}$$

Choose $q = j$, $l = i$ and sum over all i, j. We obtain

$$\sum_{x\in G} \chi_\alpha(xy)\chi_\alpha(x^{-1}) = (g/n)\chi_\alpha(y) \tag{13}$$

An equivalent form of (13) is

$$\sum_{x\in G} \chi_\alpha(yx^{-1})\chi_\alpha(x) = (g/n)\chi_\alpha(y) \tag{14}$$

7. A divisibility theorem

We shall use (14) to derive a significant result for any irreducible finite subgroup of $GL(n, C)$.

Theorem IX.5. Let G be any irreducible finite subgroup of $GL(n, C)$. Then n divides the order of G.

Proof. Denote the elements of G by $x_1, x_2, \ldots, x_g$. Set

$$\tau_q = \chi_\alpha(x_q), \qquad c_{pq} = \chi_\alpha(x_p x_q^{-1}), \qquad 1 \le p, \quad q \le g$$

Then (14) becomes

$$\sum_{q=1}^{g} c_{pq}\tau_q = (g/n)\tau_p, \qquad 1 \leq p \leq g$$

The τ_q are not all 0 [since $\chi_\alpha(I) = n$, for example] and so this system has a nontrivial solution. It follows that g/n is an eigenvalue of the matrix $C = (c_{pq})$. Now the numbers c_{pq} are algebraic integers of $Q(\zeta_g)$, since they are sums of gth roots of unity. Hence the eigenvalues of C are also algebraic integers, and so g/n is an algebraic integer. But g/n is rational. Hence it is a rational integer, and the proof is concluded.

We are now very close to the proofs of the fundamental relations of representation theory, which depend on the so-called regular representation. These state that the number of pairwise inequivalent irreducible matrix representations of a finite group G is equal to the number of conjugacy classes of G, and that the sum of the squares of their degrees is equal to the order of G. Proofs of these statements along the lines outlined here may be found in [52]. We pursue the question no further, but go on now to finite integral matrix groups.

8. Integral matrix groups

We assume then that $G = \{A\}$ is a finite subgroup of $GL(n, Z)$ of order g. Then if $A \in G$, $\det(A) = \pm 1$. The matrices of G of determinant 1 are a normal subgroup of G of index 1 or 2. It is sometimes more convenient to study this subgroup, which we may denote conveniently by G^+.

9. Lemmas on traces

We are going to derive bounds for the order of G in terms of n. For this purpose we prove

Lemma IX.1. Let S be the set $\{A_1, A_2, \ldots, A_k\}$, where the elements of S are not necessarily distinct and belong to $GL(n, Z)$. Suppose that S has the property

$$A_iS = \{A_iA_1, A_iA_2, \ldots, A_iA_k\} = S$$

for all i such that $1 \leq i \leq k$. Then

$$\sum_{i=1}^{k} \operatorname{tr}(A_i) \equiv 0 \bmod k$$

Proof. Put $M = \sum_{i=1}^{k} A_i$. We must show that $\operatorname{tr}(M) \equiv 0 \bmod k$. By the closure property of S,

$$A_j M = \sum_{i=1}^{k} A_j A_i = \sum_{i=1}^{k} A_i = M, \qquad 1 \leq i \leq k$$

Summing over all j, we get that $M^2 = kM$. Thus the eigenvalues of M are 0 and k, each with their proper multiplicity. Hence $\operatorname{tr}(M)$ is an integral multiple of k, as required.

Lemma IX.2. The sum

$$\sigma_k = \sum_{A \in G} \{\operatorname{tr}(A)\}^k, \qquad k = 1, 2, \ldots$$

is always an integer divisible by g.

Proof. Let $G^{(k)} = \{A^{(k)}\}$, where $A^{(k)}$ represents the kth Kronecker power of A. Then the distinct elements of $G^{(k)}$ form a subgroup of $GL(n^k, Z)$ whose order divides g, since the mapping $A \longrightarrow A^{(k)}$, $A \in G$, is a homomorphism. Furthermore

$$\operatorname{tr}\{A^{(k)}\} = \{\operatorname{tr}(A)\}^k$$

The previous lemma now implies that σ_k is divisible by g, since the set $\{A^{(k)}\}$ obviously possesses the required closure property.

We note that $\sum_{A \in G} 1 = g$ is also divisible by g. Lemma IX.2 at once implies the following useful result:

Lemma IX.3. Let $f(\lambda)$ be any polynomial over Z. Then

$$\sum_{A \in G} f(\operatorname{tr}(A)) \equiv 0 \bmod g$$

We also require

Lemma IX.4. The only element of G with trace n is the identity.

Proof. Let $A \in G$, $\operatorname{tr}(A) = n$. Since $A^g = I$, the eigenvalues of A are roots of unity and therefore of absolute value 1. The sum of n numbers of absolute value 1 can be n if and only if each of them is 1. Thus all the eigenvalues of A are 1. This implies that all the eigenvalues of $B =$

$I + A + \cdots + A^{g-1}$ are g, so that B is nonsingular. But $(A - I)B = 0$, and so $A = I$. This concludes the proof.

10. A bound for the order of a finite integral matrix group

We now combine these lemmas to prove

Theorem IX.6. Let $t_1 = n, t_2, \ldots, t_r$ be the distinct values assumed by $\mathrm{tr}(A)$ as A runs over G. Then

$$(n - t_2)(n - t_3) \cdots (n - t_r) \equiv 0 \bmod g, \tag{15}$$

$$(2n)! \equiv 0 \bmod g \tag{16}$$

Proof. The polynomial

$$f(\lambda) = (\lambda - t_2)(\lambda - t_3) \cdots (\lambda - t_r)$$

is an integral polynomial which has the property that $f(\mathrm{tr}(A))$ is different from 0 if and only if $\mathrm{tr}(A) = n$; hence if and only if $A = I$, by Lemma IX.4. Thus

$$\sum_{A \in G} f(\mathrm{tr}(A)) = (n - t_2)(n - t_3) \cdots (n - t_r)$$

Congruence (15) is now a consequence of Lemma IX.3. To prove (16), note that $\mathrm{tr}(A)$ is an integer such that $|\mathrm{tr}(A)| \leq n$, since it is the sum of n gth roots of unity. Thus the only values possible for $\mathrm{tr}(A)$ are

$$0, \pm 1, \ldots, \pm n$$

It follows that $(n - t_2)(n - t_3) \cdots (n - t_r)$ is a divisor of $\prod_{i=0}^{n-1} (n - i) \cdot \prod_{i=1}^{n} (n + i) = (2n)!$ This completes the proof.

11. Congruential results of Minkowski

We now turn to some work of Minkowski, which was motivated by the problem of determining the automorphs of a positive definite quadratic form.

We first prove

Lemma IX.5. Let p be a prime, and let m be an integer > 2. Let $A \in Z_n$ satisfy

$$A^p = I, \tag{17}$$

and

(18) $$A \equiv I \bmod m$$

Then $A = I$.

Proof. Suppose that $A \neq I$. Then because of (18) we may write $A = I + mE$, where $E \in Z_n$, $E \neq 0$. We may assume that the greatest common divisor δ of the elements of E is 1, since otherwise m can be replaced by $m\delta$. From (17) we have that $(I + mE)^p = I$, which implies that

$$pE + \binom{p}{2} mE^2 \equiv 0 \bmod m^2$$

Thus $pE \equiv 0 \bmod m$, and since $\delta = 1$, $p \equiv 0 \bmod m$. This implies that $m = p$, since $m > 2$. But now

$$pE + \frac{p^2(p-1)}{2} E^2 \equiv 0 \bmod p^2$$

and since p is odd, $pE \equiv 0 \bmod p^2$, $E \equiv 0 \bmod p$, a contradiction. This completes the proof.

We use this lemma to prove

Theorem IX.7. Let k, m be positive integers. Let $A \in Z_n$ satisfy

(19) $$A^k = I,$$

(20) $$A \equiv I \bmod m$$

Then if $m > 2$, $A = I$; and if $m = 2$, $A^2 = I$.

Proof. Assume first that $m > 2$. The result is certainly true for $k = 1$, since $A = I$ then. Assume that it has been proved for all k with at most $t - 1$ prime factors, $t \geq 1$; and let k have t prime factors. Write $k = pk_1$, where p is prime and k_1 has $t - 1$ prime factors. Then (19) reads $(A^{k_1})^p = I$, and powering (20), we get $A^{k_1} \equiv I \bmod m$. By the lemma, $A^{k_1} = I$. By the induction hypothesis, $A = I$. This completes the proof for $m > 2$.

Next, assume that $m = 2$. Write $k = 2^\alpha k_1$, where $\alpha \geq 0$ and k_1 is odd. Put $B = A^{2^\alpha}$. Then (19) reads $B^{k_1} = I$, and powering (20), we get $B \equiv I \bmod 2$. Suppose that $B \neq I$. Put $B = I + 2E$, where $E \in Z_n$, $E \neq 0$. If the greatest common divisor δ of the elements of E is > 1, the previous case applies (with 2 replaced by 2δ) and we deduce that

$B = I$. Suppose then that $\delta = 1$. Then as before we have

$$2k_1E + 2k_1(k_1 - 1)E^2 \equiv 0 \bmod 8,$$
$$2k_1E \equiv 0 \bmod 4,$$
$$k_1 \equiv 0 \bmod 2$$

a contradiction. Hence $B = I$ in this case as well. Thus we have shown

$$A^{2^\alpha} = I$$

If $\alpha = 0$ or 1 we are through. Suppose that $\alpha \geq 2$. We have the identity

$$0 = A^{2^\alpha} - I = (A - I)(A + I)(A^2 + I) \cdots (A^{2^{\alpha-1}} + I)$$

Since $A \equiv I \bmod 2$,

$$A^{2^r} + I \equiv 2I \bmod 4, \qquad 1 \leq r \leq \alpha - 1$$

It follows that $\frac{1}{2}(A^{2^r} + I) \in Z_n$, and that $\det(\frac{1}{2}(A^{2^r} + I))$ is odd. Thus $A^{2^r} + I$ is nonsingular, $1 \leq r \leq \alpha - 1$. Hence $(A - I)(A + I) = 0$, and the proof is complete.

12. A theorem on a natural homomorphism

The theorem we are leading up to is concerned with the natural homomorphism φ of Z onto $Z/(m)$, where m is any positive integer. This homomorphism induces a homomorphism of

$$GL(n, Z) \qquad \text{into} \qquad GL(n, Z/(m))$$

and so a homomorphism of G onto a subgrup φG of $GL(n, Z/(m))$. The important result that follows from Theorem IX.7 is the following:

Theorem IX.8. Suppose that $m > 2$. Then the kernel of φ is the identity, so that G is isomorphic to some subgroup of $GL(n, Z/(m))$. Hence the order of G divides the order of $GL(n, Z/(m))$. Similarly, the order of G^+ divides the order of $SL(n, Z/(m))$.

Proof. The kernel of φ consists of all matrices A of G such that $A \equiv I \bmod m$. Since also $A^g = I$ and $m > 2$, the result is a consequence of Theorem IX.7.

As was mentioned previously, this result of Minkowski's was motivated by his study of the automorphs of positive definite quadratic forms.

13. Automorphs

Let M be a real $n \times n$ positive definite symmetric matrix. Let x be a real $n \times 1$ vector. Then if λ is the largest eigenvalue of M and μ the smallest eigenvalue of M, so that $\lambda \geq \mu > 0$,

$$\mu x^{\mathrm{T}}x \leq x^{\mathrm{T}}Mx \leq \lambda x^{\mathrm{T}}x \tag{21}$$

A matrix $A \in GL(n, Z)$ is an *automorph* of M if

$$A^{\mathrm{T}}MA = M \tag{22}$$

The automorphs of M constitute a subgroup of $GL(n, Z)$ which we denote by $G(M)$. This group is a finite group. For if A is written as the matrix of its column vectors,

$$A = [C_1, C_2, \ldots, C_n]$$

then (22) becomes

$$(C_i^{\mathrm{T}}MC_j) = M$$

Comparing diagonal elements, we have

$$C_i^{\mathrm{T}}MC_i = m_{ii}, \qquad 1 \leq i \leq n$$

The diagonal elements m_{ii} of M are positive, since they are values assumed by the quadratic form associated with M. Then (21) implies that

$$\begin{aligned} \mu C_i^{\mathrm{T}}C_i &\leq C_i^{\mathrm{T}}MC_i = m_{ii}, \\ C_i^{\mathrm{T}}C_i &\leq m_{ii}/\mu, \qquad 1 \leq i \leq n \end{aligned} \tag{23}$$

Inequality (23) has only finitely many solutions in integral vectors C_i. Hence $G(M)$ is finite.

Conversely, let $G = \{A\}$ be any finite subgroup of $GL(n, Z)$. Then G is a subgroup of some group $G(M)$, where M is a positive-definite symmetric matrix. In fact, we may choose

$$M = \sum_{A \in G} A^{\mathrm{T}}A$$

so that M is clearly symmetric and positive definite. Furthermore if B is any element of G, then

$$B^{\mathrm{T}}MB = \sum_{A \in G} B^{\mathrm{T}}A^{\mathrm{T}}AB = \sum_{A \in G} (AB)^{\mathrm{T}}(AB) = M$$

since AB runs over G as A does.

Thus the maximal subgroups of $GL(n, Z)$ appear as the automorph groups of M for some suitable symmetric positive-definite matrix M. This approach has been very fruitful, especially in the modern theory of quadratic forms (see [10] and [58]).

14. The finite subgroups of $GL(2, Z)$ and $GL(3, Z)$

We conclude this chapter with an applcation of the derived theorems to the determination of the finite subgroups of $GL(2, Z)$ and $GL(3, Z)$. We let C_n stand for any cyclic group of order n, and D_n for the dihedral group of order $2n$. Thus D_n may be presented as the group on two generators x, y with defining relations

$$x^2 = (xy)^2 = y^n = 1$$

In addition we let T stand for the tetrahedral group of order 12, and O for the octahedral group of order 24. Thus T may be presented as the group on two generators x, y with defining relations

$$x^2 = (xy)^3 = y^3 = 1$$

and O may be presented as the group on two generators x, y with defining relations

$$x^2 = (xy)^4 = y^3 = 1$$

Suppose now that $A \in SL(2, Z)$, A of finite period. Then the eigenvalues of A are reciprocals of one another (since $\det(A) = 1$) and are algebraic integers of degree ≤ 2. Since A is of finite period, A must be similar to the diagonal matrix of its eigenvalues, say $\operatorname{diag}(\lambda, 1/\lambda)$. It follows that r is the order of A if and only if it is the order of λ. Thus λ must be a primitive rth root of unity, and so is of degree $\varphi(r)$. Hence $\varphi(r) \leq 2$, which implies that $r = 1, 2, 3, 4, 6$. This limits λ to the values ± 1, $\pm i$, $\pm \rho$, $\pm \rho^2$, where ρ is a primitive cube root of unity, and so the trace of A can only be $0, \pm 1, \pm 2$. It follows by Theorem IX.6 that the finite subgroups of $SL(2, Z)$ have orders dividing

$$(2 - 0)(2 - 1)(2 + 1)(2 + 2) = 24$$

In fact the only possible orders are 1, 2, 3, 4, 6 and any such subgroup must be conjugate over $GL(2, Z)$ to one of the cyclic groups

$$\{I\}, \{-I\}, \{U\}, \{V\}, \{-U\}$$

where $U = \left[\begin{smallmatrix} 0 & 1 \\ -1 & -1 \end{smallmatrix}\right]$, $V = \left[\begin{smallmatrix} 0 & 1 \\ -1 & 0 \end{smallmatrix}\right]$, and $V^2 = -I$, $U^3 = I$.

Now consider any finite subgroup G of $GL(2, Z)$. Then G^+ is of index 1 or 2 in G, and $G^+ = C_1, C_2, C_3, C_4$, or C_6. Since G contains C_n as a normal subgroup of index 2 if and only if it is equal to C_{2n}, $C_2 \times C_n$, or D_n, a brief analysis shows that it is only necessary to add the groups D_1, D_2, D_3, D_4, D_6 to the above list to obtain all finite subgroups of $GL(2, Z)$. The number of new groups so obtained which are not conjugate over $GL(2, Z)$ is 8, and in terms of the canonical generators x, y of D_n [which satisfy $x^2 = (xy)^2 = y^n = 1$] a complete list is the following:

$$D_1 : x \longrightarrow \begin{bmatrix} 1 & 0 \\ 0 & -1 \end{bmatrix},$$

$$D_1 : x \longrightarrow \begin{bmatrix} 0 & 1 \\ 1 & 0 \end{bmatrix},$$

$$D_2 : x \longrightarrow \begin{bmatrix} 1 & 0 \\ 0 & -1 \end{bmatrix}, \qquad y \longrightarrow -I,$$

$$D_2 : x \longrightarrow \begin{bmatrix} 0 & 1 \\ 1 & 0 \end{bmatrix}, \qquad y \longrightarrow -I,$$

$$D_3 : x \longrightarrow \begin{bmatrix} 0 & 1 \\ 1 & 0 \end{bmatrix}, \qquad y \longrightarrow U,$$

$$D_3 : x \longrightarrow \begin{bmatrix} 0 & -1 \\ -1 & 0 \end{bmatrix}, \qquad y \longrightarrow U,$$

$$D_4 : x \longrightarrow \begin{bmatrix} 0 & 1 \\ 1 & 0 \end{bmatrix}, \qquad y \longrightarrow V,$$

$$D_6 : x \longrightarrow \begin{bmatrix} 0 & 1 \\ 1 & 0 \end{bmatrix}, \qquad y \longrightarrow \begin{bmatrix} 0 & 1 \\ -1 & 1 \end{bmatrix}$$

Thus there are in all 13 mutually nonconjugate finite subgroups of $GL(2, Z)$.

A similar discussion shows that the order of a finite subgroup of $SL(3, Z)$ must also be a divisor of 24, and that an individual element of such a group must have order 1, 2, 3, 4, 6. A careful enumeration shows that the possible groups are

$$C_1, C_2, C_3, C_4, C_6$$
$$D_4, D_6, D_8, D_{12}$$

the tetrahedral group T of order 12
the octahedral group O of order 24

It can be shown that there are in all 70 mutually nonconjugate subgroups of $GL(3, Z)$.

The maximal finite subgroups of $GL(4, Z)$ have been determined by Dade, up to conjugacy. There are nine of them, and they are described as the automorph groups of suitable symmetric positive definite matrices M. See Dade's paper [7] for a complete discussion.

EXERCISES AND PROBLEMS

1. Let G be a finite integral matrix group of order g and degree n. Let p be any prime dividing g. Then $p \leq n + 1$.

Hint: Since $p \mid g$, G must contain a matrix A of period p. Since p is a prime, at least one eigenvalue λ of A must be a primitive pth root of unity. Thus $\deg \lambda = p - 1$. But also, $\deg \lambda \leq n$. Thus $p \leq n + 1$.

2. Show that the order of a finite subgroup of $GL(4, Z)$ must divide 5760.

Chapter X

Circulants

1. The matrix P and its properties

This chapter is devoted to the study of the commutative polynomial ring

$$R = Z[P]$$

where P is the $n \times n$ permutation matrix

$$(1) \qquad P = \begin{bmatrix} 0 & 1 & 0 & \cdots & 0 \\ 0 & 0 & 1 & \cdots & 0 \\ \cdot & \cdot & \cdot & & \cdot \\ \cdot & \cdot & \cdot & & \cdot \\ \cdot & \cdot & \cdot & & \cdot \\ 0 & 0 & 0 & \cdots & 1 \\ 1 & 0 & 0 & \cdots & 0 \end{bmatrix}$$

As a permutation, P is the full n-cycle, and together with $\left[\begin{smallmatrix} 0 & 1 \\ 1 & 0 \end{smallmatrix}\right] \dot{+} I_{n-2}$ is one of the generators of the group of $n \times n$ permutation matrices.

The matrix P and its powers provides a faithful representation of degree n of the cyclic group of order n, which contains every irreducible representation (which are all of degree 1) precisely once. P is the companion matrix of the polynomial $\lambda^n - 1$, and hence nonderogatory. We have that $\lambda^n - 1 = \prod_{d|n} \Phi_d(\lambda)$, where $\Phi_d(\lambda)$ is the dth cyclotomic polynomial whose roots are the $\varphi(d)$ primitive dth roots of unity. Then the elementary divisors of $\lambda I - P$ over Q are the polynomials $\Phi_d(\lambda)$, $d \mid n$. P is also an orthogonal matrix, and hence unitarily equivalent to the diagonal matrix of its eigenvalues

$$D = \operatorname{diag}(1, \zeta, \ldots, \zeta^{n-1}) \tag{2}$$

where ζ is a primitive nth root of unity.

The matrix

$$S = (\zeta^{(i-1)(j-1)})$$

which satisfies $SS^* = S^*S = nI$ [so that $U = (1/\sqrt{n})S$ is unitary] diagonalizes P, and is of interest in number theory for other reasons. It was introduced by Schur in his proof of the formula for the Gauss sums.

We have that $U^*PU = D$. Thus if $C = f(P)$ is any polynomial in P with arbitrary real or complex coefficients, then

$$U^*CU = f(D) = \operatorname{diag}(f(1), f(\zeta), \ldots, f(\zeta^{n-1}))$$

The matrices C are known as *circulants*. Since P is nonderogatory any matrix that commutes with P must be a polynomial in P, and hence a circulant. C is a circulant if and only if C^{T} is a circulant, since $P^{\mathrm{T}} = P^{-1} = P^{n-1}$.

2. Circulants of finite period

We first determine the elements of $Z[P]$ of finite period.

Theorem X.1 The elements of $Z[P]$ of finite period are the $2n$ matrices

$$\pm P^k, \qquad 0 \le k \le n-1 \tag{3}$$

Proof. Suppose that $C \in Z[P]$, $C^e = I$. Then also $C^{\mathrm{T}} \in Z[P]$, $(C^{\mathrm{T}})^e = I$. It follows that $(CC^{\mathrm{T}})^e = I$. But the matrix CC^{T} is positive definite. Thus its eigenvalues must all be 1, which implies that $CC^{\mathrm{T}} = I$. It follows that C is a generalized permutation matrix. But the only generalized permutation matrices which are also circulants are the matrices (3), all of which are of finite period. This completes the proof.

3. The units of $R = Z[P]$

We are going to study the group of units of R, which we denote by R' to conform with previous notation. Then R' consists of all $n \times n$ rational integral unimodular circulants. We prove

Lemma X.1. R' is a finitely generated abelian group.

Proof. Certainly R' is abelian. Let $C = f(P)$ be any element of R'. Then

$$U^*CU = f(D) = \operatorname{diag}(f(1), f(\zeta), \ldots, f(\zeta^{n-1}))$$

where $f(1), f(\zeta), \ldots, f(\zeta^{n-1})$ are integers of the cyclotomic field $Q(\zeta)$ whose product is ± 1; and hence units of $Q(\zeta)$. Thus $U^*R'U$ is a subgroup of the group K of all diagonal matrices $\operatorname{diag}(\epsilon_1, \epsilon_2, \ldots, \epsilon_n)$, where $\epsilon_1, \epsilon_2, \ldots, \epsilon_n$ are units of $Q(\zeta)$. Since the group of units of $Q(\zeta)$ is finitely generated by the Dirichlet unit theorem, K must also be finitely generated. The conclusion now follows, since any subgroup of a finitely generated abelian group is finitely generated.

4. The free rank of R'

Let us define the *free rank* of a finitely generated abelian group G as the rank of the torsion-free part of G. In the next theorem we determine the free rank of R'. We will use the facts that the free rank of a subgroup of G does not exceed the free rank of G, and that the Dirichlet unit theorem implies that if $n > 2$, then the free rank of the units of $Q(\zeta)$ is $\frac{1}{2}\varphi(n) - 1$, which we write as $[(\varphi(n) - 1)/2]$ to include the cases $n = 1$ and 2.

Theorem X.2. The free rank of R' is given by

$$b(n) = 1 + [n/2] - \sigma_0(n) \tag{4}$$

where $\sigma_0(n)$ denotes the number of divisors of n.

Proof. It is sufficient to show that the free rank of $U^*R'U$ is given by (4). Let d be any divisor of n, and put

$$\zeta_d = e^{2\pi i/d}, \qquad t = \varphi(d)$$

Let $\zeta_d^{r_1}, \zeta_d^{r_2}, \ldots, \zeta_d^{r_t}$ be the roots of $\Phi_d(\lambda)$, so that these are the $\varphi(d)$

primitive dth roots of unity. Here $r_1, r_2, \ldots, r_t$ are a reduced set of residues modulo d. Define

$$E_d = \operatorname{diag}(\zeta_d^{r_1}, \zeta_d^{r_2}, \ldots, \zeta_d^{r_t})$$

Then D is similar to the direct sum

$$E = \sum_{d|n}{}^{\cdot} E_d$$

Thus it is sufficient to consider the group G of unimodular matrices of $Z[E]$. Let

$$f(E) = \sum_{d|n}{}^{\cdot} f(E_d)$$

be any element of G. Then

$$f(E_d) = \operatorname{diag}(f(\zeta_d^{r_1}), f(\zeta_d^{r_2}), \ldots, f(\zeta_d^{r_t}))$$

where $f(\zeta_d^{r_k})$, $1 \leq k \leq t$, is a unit of the cyclotomic field $Q(\zeta_d)$, since E_d is unimodular. Now let

$$\epsilon_{d,k}(\zeta_d), \qquad 1 \leq k \leq [(t-1)/2]$$

be a basis for the units of infinite order of $Q(\zeta_d)$. (If $d = 1, 2, 3, 4$, or 6, there are none, in which case the statement is vacuous.) Then clearly,

$$\epsilon_{d,k}(E_d) = \operatorname{diag}(\epsilon_{d,k}(\zeta_d^{r_1}), \epsilon_{d,k}(\zeta_d^{r_2}), \ldots, \epsilon_{d,k}(\zeta_d^{r_t})), \qquad 1 \leq k \leq [(t-1)/2]$$

is a basis for the elements of infinite order of G_d, the group of unimodular matrices of $Z[E_d]$, since the element $\epsilon_{d,k}(\zeta_d^{r_i})$ is obtained from $\epsilon_{d,k}(\zeta_d)$ by the field automorphism

$$\zeta_d \longrightarrow \zeta_d^{r_i}, \qquad 1 \leq i \leq t$$

It follows that the free rank of G_d is exactly $[(t-1)/2]$.

We now determine a set of elements which generate G (we do not maintain that they actually belong to G, although the construction can be modified to guarantee this). For each $d \mid n$, and for each k such that $1 \leq k \leq [(t-1)/2]$, put

$$F_{d,k} = \sum_{\delta|n}{}^{\cdot} f_{d,k}(E_\delta)$$

where

$$f_{d,k}(E_\delta) = \begin{cases} I, & \delta \neq d \\ \epsilon_{d,k}(E_d), & \delta = d \end{cases}$$

Clearly, G is contained in the group generated by the matrices $F_{d,k}$, and so the free rank of G does not exceed

$$\sum_{d|n}\left[\frac{\varphi(d)-1}{2}\right] = 1 + \left[\frac{n}{2}\right] - \sigma_0(n)$$

and a brief argument shows that it cannot be less, since otherwise the free rank of the units of $Q(\zeta_d)$ would be less than $\frac{1}{2}[\varphi(d) - 1)]$ for at least one divisor d of n. This completes the proof.

5. Some consequences

The following lists $b(n)$ for small values of n:

(5)

n	1	2	3	4	5	6	7	8	9	10	11	12	13	14	15	16
$b(n)$	0	0	0	0	1	0	2	1	2	2	4	1	5	4	4	4

One or two interesting corollaries are deducible from this list. Since $b(n) = 0$ for $n = 1, 2, 3, 4, 6$ [which are the values for which $Q(\zeta_n)$ has no units of infinite order] we get

Corollary X.1. If $n = 1, 2, 3, 4, 6$, then R' consists of the elements

$$\pm P^k, \qquad 0 \leq k \leq n-1$$

This observation was first made in [56]. Another consequence of the analysis is that a basis for the torsion-free part of R' may be deduced if the corresponding basis for the units of infinite order of $Q(\zeta_n)$ is known. This is particularly simple for $n = 5, 8, 12$, when $b(n)$ is 1, and was done in [69] and [1]. For $n = 5$, for example, the torsion-free part of R' is cyclic and is generated by $I - P^2 - P^3$. From this we get the interesting conclusion that the 5×5 integral unimodular circulants consist precisely of

$$\pm P^k(I - P^2 - P^3)^l$$

where $0 \leq k \leq 4$ and l is any integer. Further results of this type may be found in [1].

6. Some subgroups of R' and their ranks

We go on to consider the following subgroups of R':

The set R_1 of symmetric elements of R'.
The set R_2 of positive definite symmetric elements of R'.

The set R_3 consisting of all elements CC^T, where C runs over R'.

Then

$$R' \supset R_1 \supset R_2 \supset R_3$$

The eigenvalues of an element of R_1 are units of $Q(\zeta_n + 1/\zeta_n)$, which is of degree $\frac{1}{2}\varphi(n)$ over Q and totally real. The rank of the units of infinite order of $Q(\zeta_n + 1/\zeta_n)$ is thus $\frac{1}{2}\varphi(n) - 1$ for $n > 2$, by Dirichlet's unit theorem. It then follows exactly as in the proof of Theorem X.2 that the free rank of R_1 is also $b(n)$. Also, if $\epsilon_1, \epsilon_2, \ldots, \epsilon_t$ $t = \frac{1}{2}\varphi(n) - 1$, are generators of the units of infinite order of $Q(\zeta_n + 1/\zeta_n)$, then $\epsilon_1^2, \epsilon_2^2, \ldots, \epsilon_t^2$ are independent totally positive units of R_1. Since the eigenvalues of R_2 are totally positive units of R_1, this easily implies that the rank of R_2 is $b(n)$. Summarizing, we have the following theorem:

Theorem X.3. The free rank of R_1 and the rank of R_2 are also given by

(6) $$b(n) = 1 + [n/2] - \sigma_0(n)$$

7. Relations among R_1, R_2, R_3

The group R_3 is connected in a very simple way with R_1, at least when n is odd. We have

Lemma X.2. Let C be any element of R'. Then there is an integer k such that

$$C^T = \pm P^k C$$

If n is odd, there is an integer l such that

$$C = \pm P^l S, \qquad S \text{ symmetric}$$

consequently

(7) $$R_3 = R_1^2$$

where R_1^2 is the subgroup of R_1 whose elements are the squares of the elements of R_1.

Proof. Consider $M = C^{-1}C^T$. Since C and C^T commute, $MM^T = I$. Thus as in the proof of Theorem X.1, M must be of the form $\pm P^k$ for some integer k.

Now suppose that n is odd. Then k may be chosen even, since it may be replaced by $n + k$ if it is odd. Put $k = -2l$. Then the matrix $S = \pm P^{-l}C$ is symmetric. It follows that $CC^{\mathrm{T}} = S^2$, which implies that $R_3 = R_1{}^2$. This completes the proof.

The only elements of finite period of R_1 are $\pm I$, while R_2 is torsion-free. Since the rank of R_2 also equals $b(n)$, the table (5) shows that R_2 consists of the identity alone for $n = 1, 2, 3, 4, 6$. The remarks following Corollary X.1 show that R_2 is cyclic for $n = 5$ with generator $(I - P^2 - P^3)^2$. Similarly, it was shown in [1] that for $n = 7$, R_2, which is of rank 2, is generated by $(I - P^2 - P^5)^2$ and $(3I - P - 3P^2 + 2P^3 + 2P^4 - 3P^5 - P^6)^2$. We collect these observations into a lemma for later use:

Lemma X.3. If $n \leq 7$, $R_2 = R_1{}^2$.

8. Definition of k_n

We are going to consider the problem of classifying the elements of R_2 with respect to congruence transformations from R'. The discussion applies equally well to the elements of R_1, but the interest of the subject lies primarily in positive definite matrices. We define

$k_n = k_n(R_2, R') =$ the number of congruence classes of elements of R_2 with respect to congruence transformations from R'

9. The group-theoretic meaning of k_n

Our first result gives a group-theoretic meaning to k_n.

Theorem X.4. k_n is just $(R_2 : R_3)$, the index of R_3 in R_2.

Proof. Let

$$R_2 = \sum_k A_k R_3$$

be a coset decomposition for R_2 modulo R_3 (since all the groups involved are abelian, we need not differentiate between left and right coset decompositions). Let A be any element of R_2. Then for some k, $A = A_k B$, where B belongs to R_3. Thus B is of the form CC^{T}, where C

belongs to R'. It follows that

$$A = A_k B = C A_k C^{\mathrm{T}}$$

so that A is congruent to A_k.

Conversely, no two distinct A_k are congruent. For if $A_k = CA_l C^{\mathrm{T}}$, $C \in R'$, then $A_l^{-1} A_k = CC^{\mathrm{T}} \in R_3$, so that $l = k$. This completes the proof.

10. A bound for k_n

We next prove

Theorem X.5. Let r be the rank of R_2, so that $r = b(n)$. Then $k_n \leq 2^r$.

Proof. Let $C_1, C_2, \ldots, C_r$ be a basis for R_2. Then if C is any element of R_2, C may be written as

$$C = C_1^{e_1} C_2^{e_2} \cdots C_r^{e_r}$$

where $e_1, e_2, \ldots, e_r$ are rational integers. Write

$$e_i = 2f_i + g_i, \qquad 1 \leq i \leq r$$

where $g_i = 0$ or 1; and put

$$E = C_1^{f_1} C_2^{f_2} \cdots C_r^{f_r}$$

Then, because R_2 is abelian and the elements of R_2 are symmetric,

$$\begin{aligned} C &= E^2 C_1^{g_1} C_2^{g_2} \cdots C_r^{g_r} \\ &= E C_1^{g_1} C_2^{g_2} \cdots C_r^{g_r} E^{\mathrm{T}} \end{aligned}$$

Thus every element is congruent to one of $C_1^{g_1} C_2^{g_2} \cdots C_r^{g_r}$, and so the total number of classes does not exceed 2^r. This completes the proof.

11. Proof that k_n is a power of 2

Finally, we prove

Theorem X.6. The class number k_n is a power of 2.

Proof. Let $\bar{R}_1$ be the torsion-free part of R_1. Then certainly

$$\bar{R}_1 \supset R_2 \supset R_3 \supset \bar{R}_1^{\,2}$$

Let s be the rank of $\bar{R}_1$. Then $\bar{R}_1 / \bar{R}_1^{\,2}$ is of order 2^s. It follows that the order of R_2/R_3 must divide 2^s, so that k_n is a power of 2. This completes the proof.

12. The case $n = 8$

Lemma X.3 implies that $k_n = 1$ for $n \leq 7$. This is in agreement with the fact that every positive definite matrix of determinant 1 over Z is congruent to the identity for $n \leq 7$. The case $n = 8$ is the first nontrivial case. Theorem X.5 and the table (5) imply that the number of congruence classes is either 1 or 2. We shall determine the structure of R_2 completely for $n = 8$, and show that the number is indeed 2. This can be done systematically by the method of Theorem X.2, but the following discussion (which first appeared in [57]) serves the same purpose and is entirely elementary.

Suppose that $n = 8$, and let

(8) $C = a_0 I + a_1 P + a_2 P^2 + a_3 P^3 + a_4 P^4 + a_3 P^5 + a_2 P^6 + a_1 P^7$

be any element of R_2. Then the eigenvalues of C are the numbers

$$\alpha(\zeta) = a_0 + a_1(\zeta + \zeta^7) + a_2(\zeta^2 + \zeta^6) + a_3(\zeta^3 + \zeta^5) + a_4\zeta^4$$

where ζ runs over all the 8th roots of unity. Since C is unimodular and positive definite, the numbers $\alpha(\zeta)$ are real positive units of the particular cyclotomic field to which they belong. It follows that

$$\begin{aligned}
&a_0 + 2a_1 + 2a_2 + 2a_3 + a_4 = 1 && (\zeta = 1),\\
&a_0 - 2a_1 + 2a_2 - 2a_3 + a_4 = 1 && (\zeta = -1),\\
&a_0 \qquad\quad - 2a_2 \qquad\quad + a_4 = 1 && (\zeta^2 = -1),\\
&a_0 - a_4 + (a_1 - a_3)(\zeta - \zeta^3) = \epsilon_1 && (\zeta^4 = -1),\\
&a_0 - a_4 - (a_1 - a_3)(\zeta - \zeta^3) = \epsilon_2 && (\zeta^4 = -1)
\end{aligned}$$

The first three equations imply that $a_2 = 0$, $a_0 + a_4 = 1$, $a_1 + a_3 = 0$. Introducing these relations and $\zeta - \zeta^3 = \pm\sqrt{2}$ into the last two equations, we get

$$(2a_0 - 1)^2 - 8a_1{}^2 = \epsilon_1\epsilon_2 = 1$$

since $(2a_0 - 1)^2 - 8a_1{}^2$ is rational. Thus if $x = 2a_0 - 1$, $y = 2a_1$, we find

Lemma X.4. The element C given by (8) belongs to R_2 if and only if

$$a_0 = \frac{1 + x}{2}, \qquad a_1 = \tfrac{1}{2}y, \qquad a_2 = 0,$$

$$a_3 = -\tfrac{1}{2}y, \qquad a_4 = \frac{1 - x}{2}$$

where $x > 0$ and $x^2 - 2y^2 = 1$.

Since the fundamental solution of the Pell equation $x^2 - 2y^2 = 1$ with $x > 0$ is given by $x = 3$, $y = 2$, this result implies

Theorem X.7. R_2 is cyclic for $n = 8$ with generator

$$C_8 = 2I + P - P^3 - P^4 - P^5 + P^7 \tag{9}$$

There remains the question of deciding whether or not the circulant C_8 is congruent to the identity over Z; that is, whether or not C_8 is of the form A^TA, where A is an 8×8 unimodular integral matrix. That it is not, can be seen by the following argument: The diagonal elements of C_8 are all even, so that the quadratic form x^TC_8x represents only even numbers. The quadratic form x^TA^TAx represents both odd and even numbers, however. Thus we have shown that $k_8 = 2$.

13. Summary of further results

A good deal of additional work has been done on the determination of k_n. O. Taussky and E. Dade showed that $k_p = 1$ for all primes $p < 100$, except $p = 29$; M. Kneser showed that $k_9 = 1$, and extended his research to arbitrary positive-definite forms of determinant 1 in [22]; R. Austing showed in [1] that $k_{10} = 1$; and the writer showed that $k_{12} = 2$. A complete determination of k_n is still lacking, however.

We leave class numbers now and turn to another topic. We will requite some information about generalized permutation matrices.

14. Generalized *n*-cycles

A *generalized n-cycle* is a matrix of the form DQ, where D is a diagonal matrix such that $D^2 = I$, and Q is a permutation matrix which is an n-cycle. Thus Q must be of the form $S^{\mathrm{T}}PS$, where S is a permutation matrix. Just as with the permutation matrices, the generalized permutation matrices can be written as products of disjoint generalized n-cycles, although not necessarily in a unique way. The cycle structure of the generalized permutation matrix DQ is the same as the cycle structure of Q, however.

It does not necessarily follow that two generalized permutation matrices with the same cycle structure and same determinant are similar, as it does for the permutation matrices. In this connection, we prove the following lemma:

Lemma X.5. Let C be a generalized n-cycle. Then C and P are similar elements of the group of generalized permutation matrices if and only if $\det(C) = \det(P)$.

Proof. The condition is certainly necessary. Assume then that $\det(C) = \det(P)$. We first choose C to be the generalized n-cycle

$$T = \begin{bmatrix} 0 & t_1 & 0 & \cdots & 0 \\ 0 & 0 & t_2 & \cdots & 0 \\ \cdot & \cdot & \cdot & & \cdot \\ \cdot & \cdot & \cdot & & \cdot \\ \cdot & \cdot & \cdot & & \cdot \\ 0 & 0 & 0 & \cdots & t_{n-1} \\ t_n & 0 & 0 & \cdots & 0 \end{bmatrix}$$

where $t_i = \pm 1$, $1 \leq i \leq n$. Then the condition $\det(T) = \det(P)$ is equivalent to $t_1 t_2 \cdots t_n = 1$. Let D be the diagonal matrix $\operatorname{diag}(d_1, d_2, \ldots, d_n)$, where $d_i = \pm 1$, $1 \leq i \leq n$. We will try to determine D so that $D^{-1}PD = T$. We have

$$D^{-1}PD = DPD = \begin{bmatrix} 0 & d_1 d_2 & 0 & \cdots & 0 \\ 0 & 0 & d_2 d_3 & \cdots & 0 \\ \cdot & \cdot & \cdot & & \cdot \\ \cdot & \cdot & \cdot & & \cdot \\ \cdot & \cdot & \cdot & & \cdot \\ 0 & 0 & 0 & \cdots & d_{n-1} d_n \\ d_n d_1 & 0 & 0 & \cdots & 0 \end{bmatrix}$$

Thus we must solve the system

$$\begin{aligned} d_1 d_2 &= t_1, \\ d_2 d_3 &= t_2, \\ &\vdots \\ d_{n-1} d_n &= t_{n-1}, \\ d_n d_1 &= t_n \end{aligned}$$

Choose $d_1 = \pm 1$ arbitrarily. Then the first $n - 1$ equations imply that

$$\begin{aligned} d_2 &= d_1 t_1, \\ d_3 &= d_1 t_1 t_2, \\ &\vdots \\ d_n &= d_1 t_1 t_2 \cdots t_{n-1} \end{aligned}$$

The last equation becomes

$$d_1^2 t_1 t_2 \cdots t_{n-1} = t_n$$

which is automatically satically satisfied, since $d_1^2 = 1$ and $t_1 t_2 \cdots t_n = 1$. Thus D is determined, and T and P are similar elements of the group of generalized permutation matrices.

Now let C be any generalized n-cycle such that $\det(C) = \det(P)$. Then $C = DQ$, where Q is an n-cycle in the customary sense, and D is diagonal. Thus for some permutation matrix R, $Q = R^T P R$, and so $C = DR^T PR$. Now $E = RDR^T$ is diagonal. Thus $C = R^T EPR$, so that C and EP are similar. Since $\det(EP) = \det(C) = \det(P)$, the previous discussion now implies that C and P are similar elements of the group of generalized permutation matrices. This completes the proof.

15. A theorem on congruence

The question we wish to consider was discussed in [56] and is the following: Suppose that C is an element of R_2 which is of the form AA^T, for some integral unimodular matrix A. Is it then necessarily of the form BB^T, where B is an integral unimodular circulant? The answer is yes, and is the content of the next theorem.

Theorem X.8. Let $C = AA^T$ be a circulant, where A is an integral unimodular matrix. Then C is also of the form BB^T, where B is an integral unimodular circulant.

Proof. Since AA^T is a circulant it commutes with P, so that $PAA^TP^T = AA^T$. Put $M = A^{-1}PA$. Then M is an integral unimodular matrix of period n, and $MM^T = I$, so that it must be a generalized permutation matrix. If we write A as the matrix of its row vectors

$$A = \begin{bmatrix} A_1 \\ A_2 \\ \cdot \\ \cdot \\ \cdot \\ A_n \end{bmatrix}$$

then the relationship $PA = AM$ implies that

$$\begin{aligned} A_2 &= A_1M, \\ A_3 &= A_2M, \\ &\cdot \\ &\cdot \\ &\cdot \\ A_n &= A_{n-1}M, \\ A_1 &= A_nM \end{aligned}$$

It follows that

$$A = \begin{bmatrix} A_1 \\ A_1M \\ \cdot \\ \cdot \\ \cdot \\ A_1M^{n-1} \end{bmatrix}$$

We now show that M is a generalized n-cycle. For assume that M is the product of at least two disjoint cycles. We shall show that this implies that $\det(A)$ is even, which is a contradiction since A is unimodular. In order to show this, it suffices to consider A modulo 2. We can then regard M as a permutation matrix. For simplicity, number the elements so that the cycles are $(12 \cdots i)$, $(i+1 \quad i+2 \cdots j)$. The first of the cycles affects only the first i columns of A, the elements in each row of this submatrix being a permutation of $a_{11}, a_{12}, \ldots, a_{ii}$. The determinant of A is therefore divisible by $\alpha = a_{11} + a_{12} + \cdots$

$+ a_{1i}$, modulo 2. Applying the same argument to the second cycle shows that $\det(A)$ is divisible by $\beta = a_{1,i+1} + a_{1,i+2} + \cdots + a_{1j}$ modulo 2, and applying the argument to the product of the cycles, we find that $\det(A)$ is also divisible by $\alpha + \beta$ modulo 2. At least one of the numbers α, β, $\alpha + \beta$, is even. This implies that A is singular modulo 2, a contradiction. Hence M cannot be the product of two or more disjoint cycles.

Since M and P are of equal determinant, the lemma tells us that there is a generalized permutation matrix Q such that $M = Q^T P Q$. Hence

$$A = \begin{bmatrix} A_1 Q^T Q \\ A_1 Q^T P Q \\ \cdot \\ \cdot \\ \cdot \\ A_1 Q^T P^{n-1} Q \end{bmatrix} = \begin{bmatrix} B_1 \\ B_1 P \\ \cdot \\ \cdot \\ \cdot \\ B_1 P^{n-1} \end{bmatrix} Q = BQ$$

where $B_1 = A_1 Q^T$, and B is clearly a circulant. Thus $AA^T = (BQ)(BQ)^T = BB^T$. This completes the proof.

This theorem has provoked a good deal of interest, and has been generalized, notably by R. C. Thompson [70]. Thompson showed that the result is valid for the group matrices associated with any solvable group. (The circulant P is the group matrix associated with a cyclic group.)

16. Further problems and a theorem on automorphs

Another interesting problem, as yet unsolved, is the following: Suppose that C and K are elements of R_2 such that $C = AKA^T$, where A is some integral unimodular matrix. Does it then follow that $C = BKB^T$, where B is an integral unimodular circulant? Theorem X.8 is the case when K is congruent to the identity. The answer is in the affirmative when n is a prime < 100 other than 29, by the results of O. Taussky and E. Dade mentioned earlier. It is also in the affirmative for $n = 8, 9, 10, 12$. Further progress on this question seems to depend on a knowledge of the automorphs of the elements of R_2. In this connection, the following is of interest:

Theorem X.9. Suppose that C is an integral unimodular matrix which is rationally congruent to I; say $C = ((1/m)H)((1/m)H)^{\mathrm{T}}$, where we may assume that m is relatively prime to the greatest common divisor of the elements of H. Then A is an automorph of C if and only if

$$A = (1/m^2)HWH^{-1} \tag{10}$$

where W is an integral matrix such that

$$WW^{\mathrm{T}} = m^4 I \tag{11}$$

$$HWH^{\mathrm{T}} = 0 \bmod m^4 \tag{12}$$

Proof. Suppose first that A is an automorph of C. Since $AHH^{\mathrm{T}}A^{\mathrm{T}} = HH^{\mathrm{T}}$, the matrix $W = m^2H^{-1}AH$ satisfies (11). Since $m^2H^{-1} = H^{\mathrm{T}}C^{-1}$ and C is unimodular, W is an integral matrix. Finally, since $HWH^{\mathrm{T}} = m^4AC$, (12) is satisfied.

Now suppose that W is an integral matrix satisfying (11) and (12). Then (11) implies that $ACA^{\mathrm{T}} = C$, where A is given by (10). Furthermore (12) implies that A is an integral matrix, since C is unimodular and $m^4A = HWH^{\mathrm{T}}C^{-1}$. This completes the proof.

Notice that if C is integrally congruent to the identity then $m = 1$, and the automorphs of $C = HH^{\mathrm{T}}$ are the matrices $A = HWH^{-1}$, where W is any generalized permutation matrix.

17. The automorphs of C_8

The first nontrivial case to which this theorem may be applied is in the determination of the automorphs of C_8, given by (9). We note that C_8 may be written as

$$C_8 = \tfrac{1}{4}H^2$$

where

$$H = H^{\mathrm{T}} = I + P + P^2 - P^3 - P^4 - P^5 + P^6 + P^7$$

Thus we must determine all integral 8×8 matrices W such that

$$WW^{\mathrm{T}} = 16I, \tag{13}$$

$$HWH \equiv 0 \bmod 16 \tag{14}$$

A good deal of numerical work is involved in this determination, and we sketch the procedure briefly. First, it may be shown that any solution of (13) is equivalent by generalized permutation matrices to one of the following:

$$W_1 = 4I_8,$$
$$W_2 = 4I_4 \dotplus 2H_4,$$
$$W_3 = 2H_4 \dotplus 2H_4,$$
$$W_4 = 4I_8 - J_8,$$
$$W_5 = \begin{bmatrix} 4I_4 - J_4 & J_4 \\ J_4 & J_4 - 4I_4 \end{bmatrix}$$

Here J_n is the $n \times n$ matrix all of whose entries are 1 (so that $J_n = I + P + P^2 + \cdots + P^{n-1}$), and $H_4 = 2I_4 - J_4$. Next, condition (14) may be shown to be equivalent to

$$(J_8 - 2I)W(J_8 - 2I) \equiv 0 \bmod 16$$

Thus any solution W is of the form

$$W = Q_1 D_1 W_k D_2 Q_2, \qquad 1 \leq k \leq 5$$

where Q_1, Q_2 are permutation matrices, D_1, D_2 are diagonal matrices such that $D_1^2 = D_2^2 = 1$, and

$$(J_8 - 2I)Q_1 D_1 W_k D_2 Q_2 (J_8 - 2I) \equiv 0 \bmod 16 \tag{15}$$

If we multiply (15) on the left by Q_1^{T} and on the right by Q_2^{T}, we find that (15) is completely equivalent to

$$(J_8 - 2I)D_1 W_k D_2 (J_8 - 2I) \equiv 0 \bmod 16$$

In this form the enumeration of the solutions involves a really small number of computations, which are easily carried out on a high-speed digital computer. This was done, and it was found that there were no solutions for $k = 2$ and 5, but solutions in all other cases. For $k = 1$, for example, the set of solutions is

$$\pm 4Q_1, \qquad 4Q_1 \begin{bmatrix} I_4 & 0 \\ 0 & -I_4 \end{bmatrix} Q_2, \qquad \pm 4Q_1 \begin{bmatrix} I_6 & 0 \\ 0 & -I_2 \end{bmatrix} Q_2$$

where Q_1, Q_2 are permutation matrices. Representative solutions for

$k = 3$ and 4 are

$$DW_3D, \qquad DW_4D, \qquad D = I_7 \dotplus -I_1$$

There are 256 generalized permutation matrices which are automorphs of C_8.

Thus C_8 is quite rich in automorphs. An interesting question suggested by this problem is the following: When is an element of R_2 rationally equivalent to the identity? This is equivalent to asking when the totally positive units of $Q(\zeta)$ are of the form $\alpha\bar{\alpha}$, where α is a number of $Q(\zeta)$.

18. A modification of Witt's theorem

We now turn to another application of C_8, for which we require the following modification of Witt's theorem:

Theorem X.10. Suppose that $[c] \dotplus A$, $[c] \dotplus B$ are symmetric positive-definite matrices of Z_n which are congruent over Z. Then A and B are congruent over Z.

Proof. Let U be a unit matrix of Z_n such that $[c] \dotplus A = U^T([c] \dotplus B)U$. Partition U as

$$U = \begin{bmatrix} p & x \\ y & V \end{bmatrix}$$

where p is an integer and $V \in Z_{n-1}$. Then

$$\begin{bmatrix} c & 0 \\ 0 & A \end{bmatrix} = \begin{bmatrix} p & y^T \\ x^T & V^T \end{bmatrix} \begin{bmatrix} c & 0 \\ 0 & B \end{bmatrix} \begin{bmatrix} p & x \\ y & V \end{bmatrix}$$

from which we find that

$$y^TBy = c(1 - p^2), \qquad y^TBV = -cpx, \qquad V^TBV + cx^Tx = A$$

Since $[c] \dotplus B$ is positive definite, $c > 0$ and B is positive definite. Thus p can only assume the values $0, \pm 1$. Suppose first that $p = \pm 1$. Then $y^TBy = 0$, which implies that $y = 0$. Hence $x = 0$ as well, since $-cpx = y^TBV = 0$. Thus $V^TBV = A$ and the conclusion follows, since $\det(V) = \pm 1$ and so V must be a unit matrix of Z_{n-1}. Next, suppose that $p = 0$. Then we have that

$$(V + yx)^TB(V + yx) = V^TBV + V^TByx + x^Ty^TBV + x^Ty^TByx$$

Since $p = 0$, $y^T BV = 0$, and $y^T By = c$. Thus

$$(V + yx)^T B(V + yx) = V^T BV + cx^T x = A$$

Again, $V + yx$ must be a unit matrix of Z_{n-1} because A and B are nonsingular. This completes the proof.

The theorem actually holds if c is any positive number, A and B are arbitrary positive definite hermitian matrices, but congruence is over Z,

19. Lower bounds for class numbers

We now find a lower bound for the number of congruence classes over Z of the symmetric positive definite matrices of $SL(n, Z)$. Let rC_8 denote the direct sum of r copies of C_8, and put

$$M_r = I_{n-8r} \dotplus rC_8, \qquad 0 \le r \le [n/8], \quad n \ge 8 \tag{16}$$

Then we prove

Theorem X.11. The matrices M_r given by (16) are mutually incongruent over Z. Thus the number of congruence classes over Z of the symmetric positive-definite matrices of $SL(n, Z)$ is at least $1 + [n/8]$, and so goes to infinity with n.

Proof. Let r, r' be integers such that $0 \le r, r' \le [n/8]$, and put $s = n - 8r$, $s' = n - 8r'$. Assume that $M_r \,\tilde{C}\, M_{r'}$ over Z, and that $s \ge s'$. Then by the previous theorem,

$$I_{s-s'} \dotplus rC_8 \,\tilde{C}\, r'C_8$$

over Z. Now the diagonal elements of $r'C_8$ are all even, and so the quadratic form associated with $r'C_8$ represents only even integers. But if $s - s' > 0$, then the quadratic form associated with $I_{s-s'} \dotplus rC_8$ represents both odd and even integers. Therefore $s = s'$, $r = r'$, and the theorem is proved.

EXERCISES AND PROBLEMS

1. If n is odd then the only automorphs of P are $\pm P^k$, $0 \le k \le n - 1$. What are the automorphs of P when n is even?

2. If APA^T is a circulant, where $A \in GL(n, Z)$ and n is odd, then $A = BQ$, where B is a circulant and Q is a generalized permutation matrix.

3. Let C be the companion matrix of the nth cyclotomic polynomial $\Phi_n(\lambda)$. Let G_n be the group of unimodular matrices of $Z[C]$. Prove that the free rank of G_n is $\frac{1}{2}[\varphi(n) - 1]$, and find a basis for the torsion-free part of G_5.

4. Discuss the more general problem for the group G of unimodular matrices of $Z[C]$, where C is the companion matrix of a polynomial with integer coefficients which is irreducible over the rationals.

5. Suppose that n is prime. Prove that the zero divisors of $Z[P]$ are the matrices $(I - P)C$ or aJ_n, where C is any matrix of $Z[P]$, a is any integer, and J_n is the $n \times n$ matrix all of whose entries are 1. What is the corresponding result for the case of arbitrary n?

6. If the characteristic polynomial of C is irreducible over Q, prove that the ring $Q[C]$ is a domain of integrity.

Chapter XI

Quadratic Forms

1. The set of matrices P_n, P_n'

In this chapter we resume our study of quadratic forms, which was begun in Chapter IV. We shall consider the set of real symmetric positive-definite matrices with respect to congruence transformations over Z, or the set of hermitian positive definite matrices with respect to congruence transformations over certain euclidean rings of complex numbers. We first consider the set P_n of all real $n \times n$ symmetric positive definite matrices, and its subset P_n', of all members of P_n of determinant 1.

2. An eigenvalue inequality

Let $A \in P_n$. Then the eigenvalues of A are positive. We denote the smallest eigenvalue of A by $\mu = \mu(A)$ and the largest by $\lambda = \lambda(A)$, so that $0 < \mu \leq \lambda$. Then if x is any real $n \times 1$ vector,

$$\mu x^{\mathrm{T}} x \leq x^{\mathrm{T}} A x \leq \lambda x^{\mathrm{T}} x$$

Thus if $x \neq 0$,

$$x^{\mathrm{T}} x \leq (1/\mu) x^{\mathrm{T}} A x$$

It follows that the number of integral vectors x such that $x^{\mathrm{T}}Ax \leq c$, where c is any positive constant, does not exceed the number of solutions of the inequality

$$x^{\mathrm{T}}x \leq c/\mu$$

and so is finite. This implies that A possesses an arithmetic minimum, which we denote by $m(A)$, as was done in Chapter IV. Thus

$$m(A) = \min_{\substack{x \in V_n \\ x \neq 0}} x^{\mathrm{T}}Ax$$

where V_n is the set of integral $n \times 1$ vectors. Then it follows precisely as in Chapter IV that

(1) $$m(A) \leq (\tfrac{4}{3})^{(n-1)/2}\, d^{1/n}, \qquad d = \det(A)$$

The results of Chapter VI give the bound

(2) $$m(A) \leq (4/\pi)\Gamma[1 + (n/2)]^{2/n}\, d^{1/n}, \qquad d = \det(A)$$

which is better than (1) for large n.

3. The Hermite constant γ_n

The exact nature of $d^{-1/n}m(A)$ is not yet known. Since $m(\lambda A) = \lambda m(A)$, $\lambda > 0$, we may confine our study of $m(A)$ to A belonging to P_n'. Then we define the *Hermite constant* γ_n by

(3) $$\gamma_n = \sup_{A \in P_n'} m(A)$$

Thus $\gamma_n \geq m(A)$ for any A of P_n', and because of (1),

(4) $$\gamma_n \leq (\tfrac{4}{3})^{(n-1)/2}$$

It may be proved that there is a matrix A of P_n' such that $\gamma_n = m(A)$; see [60], for example. However we do not actually require this result in what follows.

4. An inequality of Mordell

The theorem that follows is due to Mordell [35] and provides a useful inequality satisfied by γ_n.

Theorem XI.1 (Mordell). The Hermite constant γ_n satisfies

(5) $$\gamma_1 = 1, \qquad \gamma_2 = 2/\sqrt{3},$$

(6) $$\gamma_n \le \gamma_{n-1}^{(n-1)/(n-2)}, \qquad n \ge 3$$

Proof. Certainly $\gamma_1 = 1$, and $\gamma_2 = 2/\sqrt{3}$ because of (4), and because of the matrix

$$A = \frac{1}{\sqrt{3}}\begin{bmatrix} 2 & 1 \\ 1 & 2 \end{bmatrix}$$

which belongs to P_2' and satisfies $m(A) = 2/\sqrt{3}$. Suppose then that $n \ge 3$. Let A be any matrix of P_n'. Then A^{-1} also belongs to P_n'. Set $A = (a_{ij})$, $A^{-1} = (A_{ij})$. By first replacing A^{-1} by $U^{\mathrm{T}}A^{-1}U$ [so that A is replaced by $U^{-1}A(U^{-1})^{\mathrm{T}}$], where U is some suitable matrix of $GL(n, Z)$, we may assure that

$$A_{11} = m(A^{-1})$$

Next, by transformations of the type

$$\begin{bmatrix} 1 & 0 \\ 0 & V^{\mathrm{T}} \end{bmatrix} A \begin{bmatrix} 1 & 0 \\ 0 & V \end{bmatrix}, \qquad \begin{bmatrix} 1 & 0 \\ 0 & V^{-1} \end{bmatrix} A^{-1} \begin{bmatrix} 1 & 0 \\ 0 & (V^{-1})^{\mathrm{T}} \end{bmatrix}$$

where $V \in GL(n-1, Z)$, we may retain A_{11} and make $a_{22} = m(A_1)$, where A_1 is the principal minor matrix of A obtained by deleting the first row and column of A. Then $A_1 \in P_{n-1}$ and $\det(A_1) = A_{11}$. Thus

$$a_{22} \le \gamma_{n-1} A_{11}^{1/(n-1)} = \gamma_{n-1}\{m(A^{-1})\}^{1/(n-1)}$$

Since a_{22} is a value assumed by A for a vector of V_n, we must also have

$$m(A) \le a_{22}$$

It follows that

(7) $$m(A) \le \gamma_{n-1}\{m(A^{-1})\}^{1/(n-1)}$$

Reversing the roles of A and A^{-1} we also have

(8) $$m(A^{-1}) \le \gamma_{n-1}\{m(A)\}^{1/(n-1)}$$

(7) and (8) together imply that

(9) $$m(A) \le \gamma_{n-1}^{(n-1)/(n-2)}$$

Since (9) holds for any element of P_n', we may take the supremum of both sides for all A in P_n'. The result is that

$$\gamma_n \le \gamma_{n-1}^{(n-1)/(n-2)}$$

which is what we wished to prove. This concludes the proof.

5. Further inequalities for γ_n

An inequality in the other direction was given in [42], and is as follows:

Theorem XI.2. The Hermite constant γ_n satisfies

$$\gamma_{m+n}^{m+n} \geq \gamma_m{}^m \gamma_n{}^n \tag{10}$$

and thus

$$\gamma_n \geq \gamma_{n-1}^{(n-1)/n} \tag{11}$$

Proof. We use the fact that if $A \in P_n$, $B \in P_m$, then

$$m(A \dotplus B) = \min(m(A), m(B))$$

Choose $\epsilon > 0$ arbitrarily. Then there are matrices A, B such that $A \in P_n$, $B \in P_m$, and

$$\det(A) = 1, \qquad m(A) \geq \gamma_n - \epsilon,$$
$$\det(B) = (\gamma_n/\gamma_m)^m, \qquad m(B) \geq \gamma_m(\det(B))^{1/m} - \epsilon$$

Thus $m(B) \geq \gamma_n - \epsilon$ also. Hence

$$m(A \dotplus B) = \min(m(A), \quad m(B)) \geq \gamma_n - \epsilon$$

But also

$$m(A \dotplus B) \leq \gamma_{m+n}(\det(A \dotplus B))^{1/(m+n)} = \gamma_{m+n}(\gamma_n/\gamma_m)^{m/(m+n)}$$

It follows that

$$\gamma_{m+n}(\gamma_n/\gamma_m)^{m/(m+n)} \geq \gamma_n - \epsilon \tag{12}$$

Since (12) holds for all positive ϵ, we have

$$\gamma_{m+n}(\gamma_n/\gamma_m)^{m/(m+n)} \geq \gamma_n$$

which is just (10); (11) follows from (10) by choosing $m = 1$ and replacing n by $n - 1$. This completes the proof.

Notice that

$$\gamma_n \geq 1 \tag{13}$$

since $m(I) = 1$ and $I \in P_n'$.

Formula (2) shows that $\gamma_n < Cn$, for some suitable positive constant C. Minkowski showed by the geometry of numbers that also $\gamma_n > cn$, for some other constant c. Thus

$$\log \gamma_n / \log n \longrightarrow 1 \quad \text{as} \quad n \longrightarrow \infty$$

Whether γ_n/n approaches a limit as n approaches ∞ is unknown. The actual values of γ_n are known for $1 \leq n \leq 8$, and are all of the form $2^a 3^b$, where a, b are rationals.

6. A factorization of an arbitrary complex matrix

We are now going to develop bounds for the number of congruence classes over Z of the $n \times n$ symmetric positive definite matrices over Z. For this purpose we require some elementary information about the matrices of P_n. We first prove

Lemma XI.1. Let $A = (a_{ij})$ be any $n \times n$ complex matrix. Then A may be written as

$$A = UT$$

where U is unitary and T is upper triangular. If A is real, U may be chosen orthogonal and T real.

Proof. Let $C_1 = [a_{11}a_{21} \cdots a_{n1}]^{\mathrm{T}}$ be the first column of A. If $C_1 = 0$, choose $U_1 = I$. If $C_1 \neq 0$, choose U_1 as a unitary matrix whose first column is $(1/c)C_1$, where

$$\sum_{i=1}^{n} |a_{i1}|^2 = c^2$$

Then in either case, the entries of the first column of $U_1{}^*A$ are all 0, except possibly for the (1, 1) element.

This construction may now be repeated using operations of the form

$$I_1 \dotplus U_2, I_2 \dotplus U_3, \ldots, I_{n-2} \dotplus U_{n-1}$$

where $U_2, U_3, \ldots, U_{n-1}$ are unitary matrices, until A is reduced to upper triangular form. Furthermore it is clear that if A is real, each U_i may be chosen real, so that U may be chosen orthogonal, whence T must also be real. This completes the proof.

7. Factorization of hermitian positive definite matrices

Now let A be any element of P_n. By the results of Chapter V, we may write

$$A = U^{\mathrm{T}}DU$$

where U is orthogonal and $D = \mathrm{diag}(d_1, d_2, \ldots, d_n)$, where $d_i > 0$.

Let e_i be the positive square root of d_i and put $E = \operatorname{diag}(e_1, e_2, \ldots, e_n)$, $B = U^{\mathrm{T}}EU$. Then $B \in P_n$, and $A = B^2$. Thus we have shown that every element of P_n has a square root in P_n.

Now by Lemma XI.1, B may be written as $B = VT$, where V is orthogonal and T is upper triangular. Then

$$A = B^2 = B^{\mathrm{T}}B = T^{\mathrm{T}}T$$

Thus we have also shown that A may be written as $T^{\mathrm{T}}T$, where T is real and upper triangular. We may clearly choose the diagonal elements of T positive.

We will need to know these diagonal elements in terms of the elements of A. For each r such that $1 \leq r \leq n - 1$, write T in partitioned form as

$$T = \begin{bmatrix} U_r & V_r \\ 0 & W_r \end{bmatrix}$$

where U_r is an $r \times r$ upper triangular matrix, V_r is $r \times (n - r)$, and W_r is an $(n - r) \times (n - r)$ upper triangular matrix. Then

$$A = T^{\mathrm{T}}T = \begin{bmatrix} U_r^{\mathrm{T}}U_r & * \\ * & * \end{bmatrix}$$

Let A_r be the principal minor matrix of A obtained by striking out the last $n - r$ rows and columns of A. Then

$$A_r = U_r^{\mathrm{T}}U_r$$

It follows that if

$$d_r = \det(A_r), \qquad 1 \leq r \leq n - 1$$

and if the diagonal elements of T are

$$t_r, \qquad 1 \leq r \leq n$$

then

$$d_r = t_1^2 t_2^2 \cdots t_r^2, \qquad 1 \leq r \leq n - 1$$

Thus if we define

$$d_0 = 1, \qquad d_n = \det(A)$$

then

$$t_r^2 = d_r/d_{r-1}, \qquad 1 \leq r \leq n$$

We can modify this construction so that T is lower triangular instead of upper triangular, with corresponding results about the diagonal elements of T. We collect this information into a lemma:

Lemma XI.2. Let A be any element of P_n. Then A may be written as B^2, where B also belongs to P_n, or as $T^{\mathrm{T}}T$, where T where T is lower triangular with positive diagonal elements $t_1, t_2, \ldots, t_n$, whose values are given by

$$t_r^2 = d_{r-1}/d_r, \qquad 1 \leq r \leq n \tag{14}$$

Here $d_0 = \det(A)$, $d_n = 1$, and $d_r = \det(A_r)$ for $1 \leq r \leq n-1$, where A_r is the principal minor matrix of A obtained by striking out the first r rows and columns of A.

We shall retain the notation of this lemma in the following discussion, except that we write $d = d_0 = \det(A)$.

8. Bounds for cofactors

The next lemma is actually implicit in the proof of Theorem XI.1.

Lemma XI.3. Let A belong to P_n. Then after a suitable congruence transformation, d_1 satisfies

$$d_1 \leq \gamma_n d^{(n-1)/n} \tag{15}$$

Proof. We apply a congruence transformation to A^{-1} so that the (1, 1) element of the resulting matrix is its arithmetic minimum, and make the corresponding congruence transformation on A. We continue to denote the resulting matrix by A. Since the (1, 1) element of A^{-1} is d_1/d, and the determinant of A^{-1} is $1/d$, we have

$$d_1/d = m(A^{-1}) \leq \gamma_n (1/d)^{1/n},$$
$$d_1 \leq \gamma_n d^{1-1/n}$$

Lemma XI.4. A congruence transformation of the form $I_r \dotplus U_r$, $U_r \in GL(n-r, Z)$, does not change d_s, $0 \leq s \leq r$.

Proof. If $0 \leq s \leq r$ we have

$$I_r \dotplus U_r = I_s \dotplus I_{r-s} \dotplus U_r = I_s \dotplus V_s$$

where $V_s \in GL(n-s, Z)$. Then the congruence transformation

$$(I_r \dotplus U_r)^{\mathrm{T}} A\, (I_r \dotplus U_r)$$

replaces A_s by $V_s^{\mathrm{T}} A_s V_s$, and so does not change $d_s = \det(A_s)$.

We now obtain the following theorem from these lemmas, which provides bounds for the numbers d_r, $0 \leq r \leq n$:

Theorem XI.3. After a suitable congruence transformation the numbers d_r satisfy

$$d_{n-r} \leq \gamma_{r+1}\, d_{n-r-1}^{r/(r+1)}, \qquad 1 \leq r \leq n-1, \tag{16}$$

$$d_{n-1} \leq \left\{\prod_{k=r}^{n-1} \gamma_{k+1}^{1/k}\right\}^r d^{r/n}, \qquad 1 \leq r \leq n-1 \tag{17}$$

Proof. The validity of (16) follows by induction from Lemmas XI.3 and XI.4. Inequality (17) follows from (16) by producting the inequality

$$d_{n-r}^{1/r} \leq \gamma_{r+1}^{1/r}\, d_{n-r-1}^{1/(r+1)}$$

9. Bounds for class numbers

We now use Theorem XI.3 and Lemma XI.2 to derive bounds for the diagonal elements of A, after the congruence transformations described above have been applied. We have

$$A = T^{\mathrm{T}}T$$

where T is lower triangular and has positive diagonal elements $t_1, t_2, \ldots, t_n$ satisfying (14). As in the reduction of a matrix to Hermite normal form, we can find a lower triangular matrix $U \in GL(n, Z)$ such that the congruence transformation

$$U^{\mathrm{T}}AU = (TU)^{\mathrm{T}}(TU)$$

leaves the diagonal elements t_i unaltered and makes

$$|t_{ij}| \leq \tfrac{1}{2}t_i, \qquad 1 \leq j < i \leq n$$

As before, we continue to denote the resulting matrices by A and T. Since

$$a_{ii} = t_i^{\,2} + \sum_{k=i+1}^{n} t_{ki}^2, \qquad 1 \leq i \leq n$$

we have the bounds

$$a_{ii} \leq \frac{d_{i-1}}{d_i} + \frac{1}{4}\sum_{k=i+1}^{n} \frac{d_{k-1}}{d_k}, \qquad 1 \leq i \leq n \tag{18}$$

Furthermore, since A is positive definite, we must have

$$a_{ij}^2 < a_{ii}a_{jj}, \qquad 1 \leq 1, \quad j \leq n. \tag{19}$$

Thus (17), (18), and (19) together imply the following:

Theorem XI.4. Let A be any $n \times n$ symmetric positive definite matrix over Z of determinant d. Then A is congruent over Z to a matrix $B = (b_{ij})$ such that

$$b_{ii} \leq d_{i-1} + \tfrac{1}{4}\sum_{k=i}^{n-1} d_k, \qquad 1 \leq i \leq n$$

where the numbers d_k satisfy the inequalities (17). Then because

$$|b_{ij}|^2 < b_{ii}b_{jj}, \qquad 1 \leq 1, \;\; j \leq n$$

the total number of possible matrices B is finite and does not exceed

$$M^n(2M - 1)^{(n^2-n)/2} < 2^{(n^2-n)/2}M^{(n^2+n)/2}$$

where

$$M = \max_{1 \leq i \leq n} \left\{d_{i-1} + \tfrac{1}{4}\sum_{k=i}^{n-1} d_k\right\}$$

This bound is not at all sharp, but it has the advantage of being both general and explicit. Thus we have given an effective criterion for deciding whether or not two $n \times n$ symmetric positive definite matrices over Z of the same determinant are congruent over Z.

10. Minkowski's proof of the finiteness of the class number

We now consider Minkowski's approach to the finiteness of the class number, which involves a significant generalization of the arithmetic minimum. Let V_n stand for the lattice of $n \times 1$ vectors over Z, and let A be any element of P_n, $d = \det(A)$. Then if $v_1, v_2, \ldots, v_r$ are any real $n \times 1$ vectors, the inequality

$$\mu^r \prod_{i=1}^{r} (v_i^{\mathrm{T}}v_i) \leq \prod_{i=1}^{r} (v_i^{\mathrm{T}}Av_i)$$

is valid, where $\mu = \mu(A)$ is the least eigenvalue of A. Thus if

$$w = w(v_1, v_2, \ldots, v_r) = \prod_{i=1}^{r} (v_i^{\mathrm{T}}Av_i)$$

and c is any positive number, then the number of solutions of

$$w \leq c$$

in vectors $v_1, v_2, \ldots, v_r$ belonging to V_n does not exceed the number of solutions of the inequality

$$\prod_{i=1}^{r} (v_i^{\mathrm{T}} v_i) \leq c/\mu^r$$

and hence is finite. It follows that if $1 \leq r \leq n$, then the problem of finding the minimum of w for a set of r linearly independent vectors belonging to V_n is a finite one, and so has a solution.

Choose $r = n$, and suppose that the minimum occurs for the linearly independent vectors $v_1, v_2, \ldots, v_n$ of V_n. Let

$$m_i = v_i^{\mathrm{T}} A v_i, \qquad 1 \leq i \leq n$$

and assume that the vectors are ordered so that

$$m_1 \geq m_2 \geq \cdots \geq m_n$$

Let V be the matrix

$$V = [v_1, v_2, \ldots, v_n]$$

Then V is a nonsingular integral matrix, and we may write $V = UH$ where $U \in GL(n, Z)$ and H is a lower triangular matrix in Hermite normal form, which we write as the matrix of its column vectors

$$H = [h_1, h_2, \ldots, h_n]$$

Then

$$v_i = Uh_i, \qquad 1 \leq i \leq n$$

and

$$m_i = h_i^{\mathrm{T}} U^{\mathrm{T}} A U h_i, \qquad 1 \leq i \leq n$$

We may replace A by

$$B = U^{\mathrm{T}} A U$$

with no change in the value of m_i or loss of generality. Then in the new formulation we have

$$m_i = h_i^{\mathrm{T}} B h_i, \qquad 1 \leq i \leq n$$

where

$$\begin{aligned}
h_1^{\mathrm{T}} &= [h_{11} \quad h_{21} \quad h_{31} \quad \cdots \quad h_{n1}], \\
h_2^{\mathrm{T}} &= [\,0 \quad h_{22} \quad h_{32} \quad \cdots \quad h_{n2}], \\
h_3^{\mathrm{T}} &= [\,0 \quad 0 \quad h_{33} \quad \cdots \quad h_{n3}], \\
&\vdots \\
h_n^{\mathrm{T}} &= [\,0 \quad 0 \quad 0 \quad \cdots \quad h_{nn}]
\end{aligned}$$

Let $V_{n,i}$ be the subset of V_n consisting of all vectors $x = [x_1, x_2, \ldots, x_n)^T$ of V_n such that x_i is the first nonzero coordinate. Then in the representation above we see that h_i has the following property: Of all vectors of $V_{n,i}$, h_i is one for which $h_i^T B h_i$ is least.

We may write B as $C^T C$, where C is a real lower triangular matrix with positive diagonal elements. Put

$$y = Cx \tag{20}$$

Then if $x \in V_{n,i}$, the first nonzero coordinate of $y = [y_1, y_2, \ldots, y_n]^T$ is y_i.

Now suppose that x is any nonzero element of V_n. Then $x \in V_{n,i}$ for some i such that $1 \leq i \leq n$. Then according to the property of h_i mentioned above,

$$x^T B x \geq m_i$$

Furthermore

$$x^T B x = y^T y = y_i^2 + y_{i+1}^2 + \cdots + y_n^2$$

so that

$$y_i^2 + y_{i+1}^2 + \cdots + y_n^2 \geq m_i$$

Now if x is any vector of V_n, put

$$q = q(x) = \frac{y_1^2}{m_1} + \frac{y_2^2}{m_2} + \cdots + \frac{y_n^2}{m_n}$$

Thus if $x \in V_{n,i}$, then

$$q(x) = \frac{y_i^2}{m_i} + \frac{y_{i+1}^2}{m_{i+1}} + \cdots + \frac{y_n^2}{m_n} \geq \frac{1}{m_i}(y_i^2 + y_{i+1}^2 + \cdots + y_n^2) \geq 1$$

That is, $q(x) \geq 1$ for every vector x of V_n other than $x = 0$.

It follows that the ellipsoid

$$q(x) \leq 1 \tag{21}$$

has no lattice points in its interior, other than $x = 0$; and hence by Minkowski's fundamental theorem of Chapter VI, its volume must be $\leq 2^n$. Now the volume of (21) in the y-space is

$$(m_1 m_2 \cdots m_n)^{1/2} \frac{\pi^{n/2}}{\Gamma(1 + n/2)}$$

and since the Jacobian of the transformation (20) is $\det(C) = d^{1/2}$, its volume in the x-space is

$$\left(\frac{m_1 m_2 \cdots m_n}{d}\right)^{1/2} \frac{\pi^{n/2}}{\Gamma(1 + n/2)}$$

Thus we obtain the inequality

$$m_1 m_2 \cdots m_n \leq \frac{2^{2n}\Gamma(1 + n/2)^2}{\pi^n} d \tag{22}$$

The diagonal elements of $H^T BH$ are $m_1, m_2, \ldots, m_n$. Then Hadamard's determinantal inequality implies that

$$\det(H) \leq \left(\frac{m_1 m_2 \cdots m_n}{d}\right)^{1/2} \leq \frac{2^n\Gamma(1 + n/2)}{\pi^{n/2}}$$

Since H is in Hermite normal form, the number of possible choices for H is bounded. Furthermore, if B is an integral matrix, the number of possible choices for $H^T BH$ is bounded, since the diagonal elements of $H^T BH$ are bounded. It follows that the number of choices for B is bounded.

We have therefore proved that every $n \times n$ integral symmetric positive-definite matrix A of fixed determinant d is congruent over Z to one of finitely many matrices B, thus providing another proof of the finiteness of the class number.

The case $n = 3$ is particularly simple. For then

$$\det(H) \leq \frac{2^3\Gamma(1 + \frac{3}{2})}{\pi^{3/2}} = \frac{6}{\pi} < 2$$

so that $\det(H) = 1$. Since H is in Hermite normal form, H must be I. Thus the product of the diagonal elements of B does not exceed

$$m_1 m_2 m_3 \leq (36/\pi^2)d < 4d$$

It is known that the constant $36/\pi^2$ may be replaced by 2.

11. Some subrings of complex numbers

We now turn to the study of certain important subrings of the complex numbers. Let D be a fixed square-free integer > 1. Define

$$\theta^2 = -D$$

and put

$$\omega = \begin{cases} \theta, & D \equiv 1, 2 \bmod 4 \\ (1 + \theta)/2, & D \equiv 3 \bmod 4 \end{cases}$$

Then the integers of $Q(\theta)$ are just the elements of $Z(\omega)$. These rings are not always principal ideal rings, and so do not properly belong to the class of rings that we have been considering. However some of them are euclidean rings, and it is these that we will ultimately consider.

12. The norm constant

We first investigate the following question: If z is a complex number, determine the value of k, defined by

$$k = \max_{z} \min_{\alpha \in Z[\omega]} |z - \alpha|^2 \tag{23}$$

The definition of k is justified, since the elements of $Z[\omega]$ are the vertices of a point lattice in the complex plane, and any complex number z lies in one of the fundametal parallelograms defined by this lattice. Thus we may assume that z is restricted to belong to the parallelogram P with vertices $0, 1, 1 + \omega, \omega$. Then we have

$$k = \max_{z \in P} \min\{|z|^2, |z - 1|^2, |z - \omega - 1|^2, |z - \omega|^2\}$$

Because of the symmetry, this is the same as

$$k = \max_{z \in T} \min\{|z|^2, |z - 1|^2, |z - \omega|^2\}$$

where T is the triangle with vertices $0, 1, \omega$.

It is not difficult to prove that the maximum is assumed when z is equidistant from $0, 1, \omega$. Hence k is the square of the radius of the circle circumscribing T, and we obtain

Lemma XI.5. The number k defined by (23) satisfies

$$k = \begin{cases} (D + 1)/4, & D \equiv 1, 2 \bmod 4 \\ (D + 1)^2/16D, & D \equiv 3 \bmod 4 \end{cases} \tag{24}$$

Thus for the first few values of D we have the table

(25)

D	1	2	3	5	6	7	10	11	13	15
k	$\frac{1}{2}$	$\frac{3}{4}$	$\frac{1}{3}$	$\frac{3}{2}$	$\frac{7}{4}$	$\frac{4}{7}$	$\frac{11}{4}$	$\frac{9}{11}$	$\frac{7}{2}$	$\frac{16}{15}$

Now let $\alpha = a + b\omega$ be any element of $Z[\omega]$. Define

$$N(\alpha) = \alpha\bar{\alpha} = \begin{cases} a^2 + Db^2, & D \equiv 1, 2 \bmod 4 \\ a^2 + ab + ((D + 1)/4)b^2, & D \equiv 3 \bmod 4 \end{cases}$$

Suppose that α, β are arbitrary elements of $Z[\omega]$, $\beta \neq 0$. According to the definition of k, we may find an element γ of $Z[\omega]$ such that

$$N(\alpha/\beta - \gamma) \leq k;$$

that is, such that

$$N(\alpha - \gamma\beta) \leq kN(\beta)$$

A glance at the table (25) and a few brief considerations now imply

Theorem XI.5. The ring $Z[\omega]$ is a proper euclidean ring with associated function $v(\alpha) = N(\alpha) = \alpha\bar{\alpha}$, $\alpha \in Z[\omega]$, if and only if

$$D = 1, 2, 3, 7, 11$$

We call k the *norm constant* of $Z[\omega]$.

13. Hermitian positive definite matrices and the generalized Hermite constant $\gamma_n(D)$

We are now prepared to study the set $\mathfrak{S}_n$ of $n \times n$ hermitian positive definite matrices A with respect to transformations of the type

$$U^*AU$$

where $U \in GL(n, Z[\omega])$. Just as in the previous discussion, we can prove that any such matrix A assumes its arithmetic minimum $m(A)$, where now the quadratic form associated with A is x^*Ax, x any $n \times 1$ vector. We can define the "generalized Hermite constant" $\gamma_n(D)$ by

$$\gamma_n(D) = \sup_{A \in \mathfrak{S}_n'} m(A)$$

where $\mathfrak{S}_n'$ is the subset of $\mathfrak{S}_n$ consisting of all matrices of determinant 1. The preceding proofs go over practically unchanged and we have

Theorem XI.6. Let $D = 1, 2, 3, 7, 11$. Then the generalized Hermite constant $\gamma_n(D)$ satisfies

$$\gamma_1(D) = 1, \qquad \gamma_2(D) = (1 - k)^{-1/2}, \tag{26}$$

$$\gamma_n(D) \leq \gamma_{n-1}(D)^{(n-1)/(n-2)}, \qquad n \geq 3 \tag{27}$$

Thus in particular

$$\gamma_n(D) \leq (1/(1 - k))^{(n-1)/2} \tag{28}$$

For the cases of interest, we have the following table:

(29)

D	1	2	3	7	11
k	$\frac{1}{2}$	$\frac{3}{4}$	$\frac{1}{3}$	$\frac{4}{7}$	$\frac{9}{11}$
$(1-k)^{-1/2}$	$\sqrt{2}$	2	$\frac{1}{2}\sqrt{6}$	$\frac{1}{3}\sqrt{21}$	$\frac{1}{2}\sqrt{22}$

From an examination of this table, we can conclude

Corollary XI.1. Let A be any 2×2 hermitian positive-definite matrix of determinant 1 over $Z[\omega]$. Then if $D = 1, 3$, or 7, A must be of the form B^*B, where B is a matrix over $Z[\omega]$.

14. Some consequences

Some interesting number-theoretic consequences of this corollary may be derived. We take $D = 3$ as an example. Let p be any prime $\neq 3$. Then it is not difficult to prove that the congruence

$$x^2 + xy + y^2 + 1 \equiv 0 \bmod p$$

always has solutions. Let (x, y) be a solution, and set $x^2 + xy + y^2 + 1 = mp$. Put

$$A = \begin{bmatrix} p & x + \omega y \\ x + \bar{\omega} y & m \end{bmatrix}$$

Then A is a hermitian positive definite matrix of determinant 1 over $Z[\omega]$, and consequently must be of the form B^*B, where B is a matrix over $Z[\omega]$. It follows that integers a, b, c, d exist such that

$$p = a^2 + ab + b^2 + c^2 + cd + d^2$$

The classical literature on quadratic forms is too extensive to comment on here in any detail. The results of this chapter do indicate a few of the important landmarks, however. The reader is referred to [8] or [20] for a more detailed treatment.

EXERCISES AND PROBLEMS

1. Work out the consequences of Corollary XI.1 for the two remaining cases.

2. Let N be an arbitrary positive number. Let S_N be the set of real 2×2 matrices whose elements do not exceed N in absolute value. Show that P_2' always contains matrices which are not congruent over Z to any element of S_N.

References

1. R. Austing, Groups of unimodular circulants. *J. Res. Nat. Bur. Standards* **69 B**, 313–318 (1965).
2. H. Bass, M. Lazard, and J.-P. Serre, Sous-groupes d'indice fini dans $SL(n, Z)$. *Bull. Amer. Math Soc.* **70**, 385–392 (1964).
3. Z. I. Borevich and I. R. Shafarevich, "Number Theory," Academic Press, New York, 1969.
4. J. W. S. Cassels, "An Introduction to the Geometry of Numbers," Springer, Berlin, 1959.
5. H. S. M. Coxeter and W. O. J. Moser, "Generators and Relations for Discrete Groups," Springer, Berlin, 1957.
6. C. W. Curtis and I. Reiner, "Representation Theory of Finite Groups and Associative Algebras," Wiley (Interscience), New York, 1962.
7. E. C. Dade, The maximal finite groups of 4×4 integral matrices. *Illinois J. Math.* **9**, 99–122 (1965).
8. L. E. Dickson, "Modern Elementary Theory of Numbers," University of Chicago Press, Chicago, 1939.
9. L. E. Dickson, "Linear Groups," Dover, New York, 1958.
10. M. Eichler, "Quadratische Formen und orthogonale Gruppen," Springer, Berlin, 1952.
11. L. R. Ford, "Automorphic Functions," McGraw-Hill, New York, 1929.
12. L. Greenberg and M. Newman, Normal subgroups of the modular group. *J. Res. Nat. Bur. Standards* **74B**, 121–123 (1970).

13. M. Hall, "The Theory of Groups," Macmillan, New York, 1959.
14. M. Hall and H. J. Ryser, Normal completions of incidence matrices. *Amer. J. Math.* **76**, 581–589 (1954).
15. G. H. Hardy and E. M. Wright, "An Introduction to the Theory of Numbers," Clarendon Press, Oxford, 1960.
16. E. Hecke, "Vorlesungen über die Theorie der algebraischen Zahlen," Leipzig, 1954.
17. C. H. Houghton, A question of M. Newman and J. R. Smart. *Duke Math. J.* **32**, 541–543 (1965).
18. L. K. Hua and I. Reiner, On the generators of the symplectic modular group. *Trans. Amer. Math. Soc.* **65**, 415–426 (1949).
19. L. K. Hua and I. Reiner, Automorphisms of the unimodular group. *Trans. Amer. Math. Soc.* **71**, 331–348 (1951).
20. B. W. Jones, "The Arithmetic Theory of Quadratic Forms," Carus Monograph No. 10, Wiley, New York, 1950.
21. Irving Kaplansky, Elementary divisors and modules. *Trans Amer. Math. Soc.* **66**, 464–491 (1949).
22. M. Kneser, Klassenzahlen definiter quadratischer Formen. *Arch. Math.* **8**, 241–250 (1958).
23. A. G. Kurosh, "The Theory of Groups," Vols. I and II, Chelsea, New York, 1955.
24. E. Landau, "Elementary Number Theory," Chelsea, New York, 1958.
25. C. G. Latimer and C. C. MacDuffee, A correspondence between classes of ideals and classes of matrices. *Ann. of Math.* **34**, 313–316 (1933).
26. J. Lehner and M. Newman, Real two-dimensional representations of the modular group and related groups. *Amer. J. Math.* **87**, 945–954 (1965).
27. J. Lehner and M. Newman, Real two-dimensional representations of the free product of two finite cyclic groups. *Proc. Cambridge Philos. Soc.* **62**, 135–141 (1966).
28. C. C. MacDuffee, "The Theory of Matrices," Chelsea, New York, 1946.
29. N. H. McCoy, "The Theory of Rings," Macmillan, New York 1964.
30. D. M. McQuillan, Classification of normal congruence subgroups of the modular group. *Amer. J. Math.* **87**, 285–296 (1965).
31. M. Marcus and H. Minc, "A Survey of Matrix Theory and Matrix Inequalities," Allyn and Bacon, Boston, 1964.
32. J. L. Mennicke, Finite factor groups of the unimodular group. *Ann. of Math.* **81**, 31–37 (1965).
33. J. L. Mennicke, Zur Theorie der Siegelschen Modulgruppe. *Math. Ann.* **159**, 115–129 (1965).
34. H. Minkowski, "Geometrie der Zahlen," Chelsea, New York, 1953.
35. L. J. Mordell, Observation on the minimum of a positive quadratic form in eight variables. *J. London Math. Soc.* **19**, 3–6 (1944).
36. T. S. Motzkin, The euclidean algorithm. *Bull. Amer. Math. Soc.* **55**, 1142–1146 (1949).
37. T. Nagell, "Introduction to Number Theory," Wiley, New York, 1951.

38. M. Newman, Structure theorems for modular subgroups. *Duke Math. J.* **22**, 25–32 (1955).
39. M. Newman, An inclusion theorem for modular groups. *Proc. Amer. Math. Soc.* **8**, 125–127 (1957).
40. M. Newman, The structure of some subgroups of the modular group. *Illinois J. Math.* **6**, 480–487 (1962).
41. M. Newman, Some free products of cyclic groups. *Michigan Math. J.* **9**, 369–373 (1962).
42. M. Newman, Bounds for cofactors and arithmetic minima of quadratic forms. *J. London Math. Soc.* **38**, 215–217 (1963).
43. M. Newman, Normal congruence subgroups of the modular group. *Amer. J. Math.* **85**, 419–427 (1963).
44. M. Newman, Normal congruence subgroups of the $t \times t$ modular group. *Bull. Amer. Math. Soc.* **69**, 619–620 (1963).
45. M. Newman, A complete description of the normal subgroups of genus 1 of the modular group. *Amer. J. Math.* **86**, 17–24 (1964).
46. M. Newman, Free subgroups and normal subgroups of the modular group. *Illinois J. Math.* **8**, 262–265 (1964).
47. M. Newman, On a problem of G. Sansone. *Ann. Mat. Pura Appl.* **(4) 65**, 27–34 (1964).
48. M. Newman, Normal subgroups of the modular group which are not congruence subgroups. *Proc. Amer. Math. Soc.* **16**, 831–832 (1965).
49. M. Newman, Classification of normal subgroups of the modular group. *Trans. Amer. Math. Soc.* **126**, 267–277 (1967).
50. M. Newman, Pairs of matrices generating discrete free groups and free products. *Michigan Math. J.* **15**, 155–160 (1968).
51. M. Newman, Maximal normal subgroups of the modular group. *Proc. Amer. Math. Soc.* **19**, 1138–1144 (1968).
52. M. Newman, "Matrix Representations of Groups," National Bureau of Standards Applied Mathematics Series 60, 1968.
53. M. Newman and I. Reiner, Inclusion theorems for congruence subgroups. *Trans. Amer. Math. Soc.* **91**, 369–379 (1959).
54. M. Newman and J. R. Smart, Modulary groups of $t \times t$ matrices. *Duke Math. J.* **30**, 253–257 (1963).
55. M. Newman and J. R. Smart, Symplectic modulary groups. *Acta Arith.* **9**, 83–89 (1964).
56. M. Newman and O. Taussky, On a generalization of the normal basis in abelian algebraic number fields. *Comm. Pure Appl. Math.* **9**, 85–91 (1956).
57. M. Newman and O. Taussky, Classes of positive definite unimodular circulants. *Canad. J. Math.* **9**, 71–73 (1956).
58. O. T. O'Meara, "Introduction to quadratic forms," Springer, Berlin, 1963.
59. H. Pollard, "The Theory of Algebraic Numbers," Carus Monograph No. 9, Wiley, New York, 1950.
60. R. A. Rankin, On positive definite quadratic forms. *J. London Math. Soc.* **28**, 309–314 (1953).

61. R. A. Rankin, "The Modular Group and Its Subgroups," The Ramanujan Institute, Madras, 1969.
62. I. Reiner, Symplectic modular complements. *Trans. Amer. Math. Soc.* **77**, 498–505 (1954).
63. I. Reiner, Normal subgroups of the unimodular group. *Illinois J. Math.* **2**, 142–144 (1958).
64. I. Reiner, Subgroups of the unimodular group. *Proc. Amer. Math. Soc.* **12**, 173–174 (1961).
65. I. Reiner and J. D. Swift Congruence subgroups of matrix groups. *Pacific J. Math.* **6**, 529–540 (1956).
66. C. L. Seigel, Symplectic geometry. Amer. J. Math. **65**, 1–86 (1943).
67. O. Taussky, On a theorem of Latimer and MacDuffee. *Canad. J. Math.* **1**, 300–302 (1949).
68. O. Taussky, Matrices of rational integers. *Bull. Amer. Math. Soc.* **66**, 327–345 (1960).
69. O. Taussky, Unimodular integral circulants. *Math. Z.* **63**, 286–289 (1955).
70. R. C. Thompson, Unimodular group matrices with rational integers as elements. *Pacific J. Math.* **14**, 719–726 (1964).
71. K. Wohlfahrt, An extension of F. Klein's level concept. *Illinois J. Math.* **8**, 529–535 (1964).
72. H. Zassenhaus, "The Theory of Groups," Chelsea, New York, 1949.

Index

P

Q

R

S

T

U

V

W

Z

Pure and Applied Mathematics

A Series of Monographs and Textbooks

1: Arnold Sommerfeld. Partial Differential Equations in Physics. 1949 (Lectures on Theoretical Physics, Volume VI)

2: Reinhold Baer. Linear Algebra and Projective Geometry. 1952

3: Herbert Busemann and Paul Kelly. Projective Geometry and Projective Metrics. 1953

4: Stefan Bergman and M. Schiffer. Kernel Functions and Elliptic Differential Equations in Mathematical Physics. 1953

5: Ralph Philip Boas, Jr. Entire Functions. 1954

6: Herbert Busemann. The Geometry of Geodesics. 1955

7: Claude Chevalley. Fundamental Concepts of Algebra. 1956

8: Sze-Tsen Hu. Homotopy Theory. 1959

9: A. M. Ostrowski. Solution of Equations and Systems of Equations. Second Edition. 1966

10: J. Dieudonné. Treatise on Analysis. Volume I, Foundations of Modern Analysis, enlarged and corrected printing, 1969. Volume II, 1970.

11: S. I. Goldberg. Curvature and Homology. 1962.

12: Sigurdur Helgason. Differential Geometry and Symmetric Spaces. 1962

13: T. H. Hildebrandt. Introduction to the Theory of Integration. 1963.

14: Shreeram Abhyankar. Local Analytic Geometry. 1964

15: Richard L. Bishop and Richard J. Crittenden. Geometry of Manifolds. 1964

16: Steven A. Gaal. Point Set Topology. 1964

17: Barry Mitchell. Theory of Categories. 1965

18: Anthony P. Morse. A Theory of Sets. 1965

19: Gustave Choquet. Topology. 1966

20: Z. I. Borevich and I. R. Shafarevich. Number Theory. 1966

21: José Luis Massera and Juan Jorge Schaffer. Linear Differential Equations and Function Spaces. 1966

22: Richard D. Schafer. An Introduction to Nonassociative Alegbras. 1966

23: Martin Eichler. Introduction to the Theory of Algebraic Numbers and Functions. 1966

24: Shreeram Abhyankar. Resolution of Singularities of Embedded Algebraic Surfaces. 1966

25: François Treves. Topological Vector Spaces, Distributions, and Kernels. 1967
26: Peter D. Lax and Ralph S. Phillips. Scattering Theory. 1967.
27: Oystein Ore. The Four Color Problem. 1967
28: Maurice Heins. Complex Function Theory. 1968
29: R. M. Blumenthal and R. K. Getoor. Markov Processes and Potential Theory. 1968
30: L. J. Mordell. Diophantine Equations. 1969
31: J. Barkley Rosser. Simplified Independence Proofs: Boolean Valued Models of Set Theory. 1969
32: William F. Donoghue, Jr. Distributions and Fourier Transforms. 1969
33: Marston Morse and Stewart S. Cairns. Critical Point Theory in Global Analysis and Differential Topology. 1969
34: Edwin Weiss. Cohomology of Groups. 1969
35: Hans Freudenthal and H. De Vries. Linear Lie Groups. 1969
36: Laszlo Fuchs. Infinite Abelian Groups: Volume I. 1970
37: Keio Nagami. Dimension Theory. 1970
38: Peter L. Duren. Theory of H^p Spaces. 1970.
39: Bodo Pareigis. Categories and Functors. 1970
40: Paul L. Butzer and Rolf J. Nessel. Fourier Analysis and Approximation: Volume 1, One-Dimensional Theory. 1971
41: Eduard Prugovečki. Quantum Mechanics in Hilbert Space. 1971
42: D. V. Widder: An Introduction to Transform Theory. 1971
43: Max D. Larsen and Paul J. McCarthy. Multiplicative Theory of Ideals. 1971
44: Ernst-August Behrens. Ring Theory. 1972
45: Morris Newman. Integral Matrices. 1972
46: Glen E. Bredon. Introduction to Compact Transformation Groups. 1972
47: Werner Greub, Stephen Halperin, and Ray Vanstone. Connections, Curvature, and Cohomology: Volume I, De Rham Cohomology of Manifolds and Vector Bundles. 1972

In preparation

Ronald G. Douglas. Banach Algebra Techniques in Operator Theory

Xia Dao-xing. Measure and Integration Theory of Infinite-Dimensional Spaces: Abstract Harmonic Analysis

E. R. Kolchin. Differential Algebra and Algebraic Groups

T. Benny Rushing. Topological Embeddings

Willard Miller, Jr. Symmetry Groups and Their Applications